The Saga of Mathematics
A Brief History

Marty Lewinter
Purchase College

William Widulski
Westchester Community College

Prentice
Hall

Prentice Hall, Upper Saddle River, New Jersey 07458

Library of Congress Cataloging-in-Publication Data
    The saga of mathematics: a brief history/Marty Lewinter,
    William Widulski.
       p. cm.
    Includes bibliographical references and index.
    ISBN 0-13-034079-0 (alk. paper)
    1. Mathematics–History. I. Widulski, William. II. Title.

QA21.L49   2002
510'.9–dc21   2001024580   CIP

Acquisition Editor: *George Lobell*
Editor-in-Chief: *Sally Yagan*
Vice President/Director of Production and Manufacturing: *David W. Riccardi*
Executive Managing Editor: *Kathleen Schiaparelli*
Senior Managing Editor: *Linda Mihatov Behrens*
Production Editor: *Barbara Mack*
Manufacturing Buyer: *Alan Fischer*
Manufacturing Manager: *Trudy Pisciotti*
Marketing Manager: *Angela Battle*
Editorial Assistant: *Melanie Van Benthuysen*
Art Director: *Jayne Conte*
Cover Design: *Bruce Kenselaar*
Cover Photo: *TASKIN. Harpsichord, signed and dated by
Taskin, 1786. Victoria & Albert Museum, London, Great Britain/Art
Resource, NY*

 © 2002 by Prentice-Hall, Inc.
Upper Saddle River, New Jersey 07458

Printed in the United States of America
10   9   8   7   6   5   4   3   2   1

ISBN 0-13-034079-0

Pearson Education LTD., *London*
Pearson Education Australia PTY, Limited, *Sydney*
Pearson Education Singapore, Pte. Ltd
Pearson Education North Asia Ltd, *Hong Kong*
Pearson Education Canada, Ltd., *Toronto*
Pearson Educacion de Mexico, S.A. de C.V.
Pearson Education - Japan, *Tokyo*
Pearson Education Malaysia, Pte. Ltd

To my mom who encouraged me to pursue mathematics and
music, and taught me to respect scholarship.
- Marty Lewinter

To my parents whose constant support and guidance
have given me opportunities I would otherwise have never had.
- William Widulski

Contents

Appendices

Preface

This book was written to accomplish several things. Firstly, it is an inexpensive text for courses in the history of math. Secondly, it is a vehicle of the authors' passion (hopefully contagious) for mathematics. Thirdly, the book brings out the relation of mathematics to music, art, science, technology, and philosophy – in short, this is an interdisciplinary adventure for readers of all ages.

Finally, it is suitable for a "general education" course in mathematics. Teaching mathematics to nonmajors poses a difficult problem across the country. This book is, we believe, a good solution. It makes mathematics relevant to many other disciplines and emphasizes the cleverness and beauty inherent in this subject. It is our hope that the reader will dine on the banquet of ideas in this text with gusto.

The material is very accessible and, we think, well motivated. It features geometry, number theory, algebra, probability, graph theory, and ancient and modern counting systems, including binary arithmetic for the computer age. The first author has taught "History of Mathematics" for a decade and, after wandering from text to text, decided to write one that addresses the needs of the reluctant, ill-prepared (perhaps even math-phobic?) college student. With a light peppering of good-natured humor, the book touches on science, navigation, commerce, the calendar, music, art, philosophy, and, of course, history. In short, it has something for everyone.

There are many worked out examples in the text. The plentiful collection of homework exercises range from easy to hard, giving the instructor some flexibility in the level of difficulty of the course. Each chapter ends with a list of suggested reading.

Chapter 1, "Oh So Mysterious Egyptian Mathematics!," begins with primitive counting by grouping. Egyptian arithmetic, geometry, and algebra are then presented in a way that will interest history and anthropology majors. The concept of unit-based measurement is introduced. The births of astronomy, religion, and the calendar are linked.

In Chapter 2, "Mesopotamia Here We Come," the Babylonian position system and its similarity to (and differences from) the present day number system are presented in a historical context. Their algorithms for division and for calculating square roots are included. Babylon cosmology is shown to have influenced the European worldview up to the seventeenth century.

Chapter 3, "Those Incredible Greeks!," is the first chapter to feature mathematicians, their lives, and their contributions. Starting with Thales, the father of "proof," the text continues with the Pythagorean contributions to number theory (perfect, deficient, and abundant numbers, for example), tiling of the plane, and Greek "music theory" – including several concrete mathematical arguments such as a proof that the angle sum of a triangle is 180 degrees and a proof of the irrationality of $\sqrt{2}$. The chapter closes with the amazing feat of Hippocrates of Chios – the squaring of the lune.

In Chapter 4, "Greeks Bearing Gifts," we investigate the relevance of mathematics to Greek philosophy, especially to the Plato versus Aristotle dispute on the nature of universals, or "forms." Euclid is one of the heroes of Hellenistic mathematics in Alexandria and his contributions to the logical foundation of geometry and to number theory are discussed at length. His proof of the infinitude of the primes and his algorithm for the GCD of two numbers are presented. The great Archimedes' many contributions come next, and the chapter closes with Ptolemy's seed of trigonometry and his map of the ancient world.

Chapter 5 is entitled "Must All Good Things Come to an End?" The Classical world gives way to the rise of Christianity and Islam. We examine the contributions of the Islamic mathematicians and the development of Hindu-Arabic arithmetic - all in a historical perspective.

In Chapter 6, "Europe Smells the Coffee," the saga continues. Fibonacci, a.k.a. Leonardo of Pisa, writes several important books, a few universities dot the map of High Middle Ages Europe, Aquinas defends reason, and Nicole Oresme studies infinite series.

The theme of Chapter 7, "Mathematics Marches On," is the influence of mathematics on music and art. This is achieved without getting too technical. Visual perspective is described. The mathematical aspects of music such as pitch, meter, duration, volume, and intervals are presented.

In Chapter 8, "A Few Good Men," we focus on Copernicus, Brahe, Kepler, Galileo, Francis Bacon, and the revolution they ushered in. Galileo's analysis of the motion of falling objects involves an ancient Greek procedure for summing the first n odd numbers. With the invention of the printing press and the coming of the Reformation, the world was ripe for the great mathematics of the seventeenth century.

The next chapter, "A Most Amazing Century of Mathematical Marvels!," is organized as follows:

1. Fermat and number theory,
2. Pascal and probability,
3. Descartes and analytic geometry, and
4. Newton and the calculus.

The material is presented in some detail. The relevance to science, technology, and commerce is emphasized.

In Chapter 10, "The Age of Euler," we launch an excursion into graph theory, number theory, and the three-dimensional space R^3.

Chapter 11, "A Century of Surprises," deals with some of the themes of the nineteenth century – vectors, non-Euclidean geometry, and field theory. There is a nice discussion of the sine function and its relevance to AM and FM radio.

The story ends with Chapter 12, "Ones and Zeros." The twentieth century saw the advent of the computer. Its language, binary arithmetic, is connected to the "doubling" employed in ancient Egyptian multiplication and to the position system of ancient Mesopotamia. The analysis in the text reinforces the student's grasp of the position system of decimal arithmetic.

We added an optional chapter, "Some More Math Before You Go," which contains a host of topics that may be called upon to bolster the mathematics content of a general education course. We include solving quadratic equations by factoring and by using the quadratic formula, graphing parabolas and circles, solving simultaneous equations, and reviewing operations with fractions.

We would like to thank the following people for their assistance and/or encouragement: Timothy Bocchi (a great teacher and a great friend); Dr. Jerome Huyler (author of *Locke in America*); Professor Frank Harary (an inspiring, world-class mathematician!); Kathy Lavelle, Bert Liberi, Joyce McQuade, and Mike Shub (for supplying

some material); and the illustrator Amy Herrmann (whose entirely
original artwork elevates the quality of the book enormously).

Marty Lewinter
mjlewin@earthlink.net

William Widulski
william.widulski@sunywcc.edu

May 1, 2001

Journey*

In the darkness of the night
When the world's devoid of sight -
Just that time when most eyes rest,
Then my vision's at its best.

Why, I can see for centuries!
Great men and deeds - great mysteries
And thoughts that set the world ablaze
Are now encompassed in my gaze.

From here within my time-machine
Ancient things - and kings - are seen!
Distant past revealed to me -
Indeed, I witness history!

Splendid battles on my stage,
Glimpses of each golden age,
When art and science bloomed and grew -
Yes each scene shows me something new,
Till tired yawns upon me creep
And make me close my book and sleep!

Chapter 1

Oh, So Mysterious Egyptian Mathematics!

Mathematics had its origins eons before recorded history. Before the rise of cities and the use of farming, some 10,000 years ago, early humans were hunter/gatherers. They hunted animals and gathered

1

berries, sharp stones, wooden branches, and so forth. Of course, modern humans hunt for bargains and gather things into their shopping baskets in giant supermarkets. While modern supermarkets use computers to add up your tab and to keep track of inventory, the mathematical needs and methods of early humans were very simple. How simple? They counted. How large is a hunting party? How many lions were hunted? How many sunrises passed since the last kill?

Their methods were simple. They used notches on a bone or a stack of pebbles. They probably had grunts to represent different small numbers, say, up to five. Anthropological evidence tells us that in some societies, *five* was given the same name as *hand*. So *ten* was called two *hands*.

Mathematics was not very abstract back then. It was visual. Fingers were used for counting. In one culture, six was called "taking the thumb" since it involved one fist for the first five things and then the thumb of the other hand. Imagine this being done in a large accounting firm today, in which we must compute 34.27% of $23,667,385.

There gradually emerged the realization that by grouping quantities, one could denote larger numbers. Thus, four stones and a pebble might represent 21, if a stone represents five and a pebble one. This was a step in the right direction. In some sense, we do this today when we take out a ten-dollar bill and three singles to pay a $13 check. Prisoners do this when they scratch groups of five marks on a wall, the fifth mark lying across the other four. Of course, lifers will eventually run out of wall space.

To summarize, simple counting requirements gave birth to simple mathematics, never progressing beyond primitive grouping schemes, using 2, 3, 5, and 10 as the grouping amount.

And then it happened! Things would never be the same. Humans discovered agriculture. This requires settlements or cities. Humans became agrarian city-dwellers, creating a need for a calendar for planting and harvesting, geometry for building things, arithmetic for trading purposes. Yes! Division of labor permitted specialization that in turn depended on trade. A builder could trade his services with a farmer who could trade his services with a tailor. Barter requires mathematics – how many pineapples for your six loaves of bread? On the other hand, production requires mathematics. How much flour would a baker need this week? Eventually, humans stumbled onto an incredible measuring idea. Set aside a fixed amount of the thing you wish to measure, and then count how many of these units there are in your pile. We do this today when we use quarts, pounds, feet, days, "miles per hour," square miles, and so on. Mathematics was in full swing!

Early societies soon accumulated a surplus of goods, enabling the creation of a priestly class. The job of this priestly class, in addition to religious pursuits, was to establish a writing system, keep records, develop a calendar, watch the skies for any ominous astrological events, and contemplate a variety of other intellectual endeavors.

It was soon convenient for nearby societies to trade, requiring ships and navigation. Early navigators of the Mediterranean looked up (at night) and did geometry on the stars – something many amateur sailors do today. Sadly, it was necessary for monarchs to have armies to maintain their rule against external and internal challenge, again requiring counting. How large is a division? a platoon? a squad? Soldiers need to be fed or paid. So do bureaucrats, hence giving rise to the need for taxation. This involves a fair bit of math – especially land assessment. In fact, farm boundaries must be defined and restored after flooding. Also, their areas must be computed for tax purposes.

One of the earliest well organized societies, Egypt, is where we begin our story. This great story tells a tale of incredible cleverness.

Ancient Egypt was ruled by a succession of pharaohs for approximately three thousand years. (By contrast, American society is a mere few hundred years old.) Their records precede 3000 B.C. and give us a glimpse of their sophistication. Let's examine their mathematics in some detail. The first steps in any discipline are often the most difficult and clever.

We will refer to whole numbers like 5 and 38 as *positive integers*, or just *integers*, when the context is clear. The Egyptians denoted the integers using the hieroglyphics shown above: the *stick* for 1, the *heel bone* for 10, the *scroll* for 100, the *lotus flower* for 1,000, the *bent finger* or *snake* for 10,000, the *burbot fish* or *tadpole* for 100,000, and the *astonished man* for 1,000,000. This enabled the Egyptians to write very large numbers describing vast quantities of food, soldiers, slaves, or livestock. Their system was one of the first recorded examples of the use of base-ten grouping. The number 243,526 for example, was written as

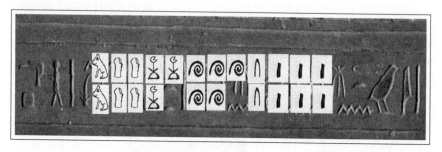

Note that *ten* of any symbol could be replaced by *one* of the next higher symbol, enabling addition to be done reasonably efficiently. In fact the genius of this idea is astonishing! Adding six "tens" and five "tens" is exactly like adding six "ones" and five "ones." We consolidate ten of the eleven symbols into the next higher symbol and are left with only one of the original symbol. This is very much like the counting of, say, ten dollar bills, which can be done like counting singles. A stack of ten singles may be exchanged for a ten-dollar bill, while a stack of ten "tens" may be exchanged for a "hundred."

An amazing feature of ancient Egyptian mathematics was their unique method of multiplication, which they correctly viewed as repeated addition, based entirely on doubling![1]

---

[1] As an economics professor once observed, "Ireland is the richest country in the world because its capital is always Dublin."

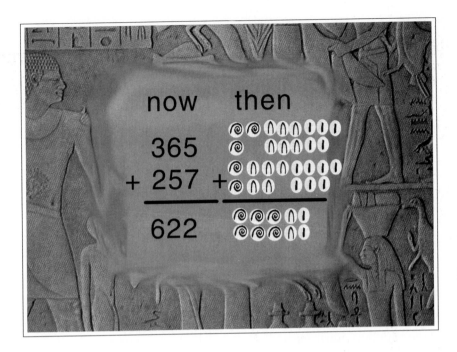

Doubling a quantity written in hieroglyphics is quite simple. One visually replaces each symbol by two, remembering to replace ten of anything by the next higher symbol (as was done in addition).

Now starting with one and doubling, they obtained a never-ending sequence of numbers 1, 2, 4, 8, 16, 32, ... which we will temporarily call "doubling numbers." Readers familiar with algebra will recognize these numbers as powers of two. They get large very quickly — the twenty-first number on the list, 2^{20}, or two to the twentieth power, exceeds one million.

The remarkable fact that the Egyptians figured out is that any integer can be written as a sum of "doubling numbers" without repeating any of them. Some examples demonstrating this are

$$11 = 1 + 2 + 8 \qquad 23 = 1 + 2 + 4 + 16 \qquad 44 = 4 + 8 + 32$$

This fact was cleverly exploited to yield the following procedure (or as mathematicians say, *algorithm*) for multiplying. Suppose, for example, that an ancient Egyptian wanted to find out how many eggs are in the seventeen dozen he just bartered at the 24-hour Pyramid Mart. He would organize his work in two columns to get 12×17. The left column starts with 1 and the right column with the larger of

the two numbers, in this case, 17. Then he would successively double these numbers until he had enough "doubling" numbers in the left column to add up to the smaller number, 12. See Figure 1-1.

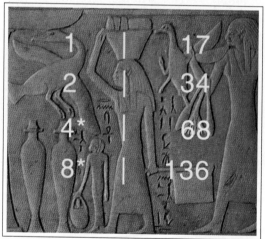

Figure 1-1: Egyptian Multiplication

He would notice that the starred numbers 4 and 8 on the left add up to 12, telling him that the numbers opposite them in the right column should be added to obtain the result. If we add 68 and 136 we get 204, which is the correct product of 12 and 17. Not bad for thousands of years ago – huh?

This Egyptian method of multiplication is also known as the *dyadic* method, meaning doubling. For another example consider,

Example 1-1 *Multiply* 20 × 24 *using the Egyptian method of multiplication.*

> *Step 1* Start with 1 and the larger number 24. Keep doubling both numbers until the left side gets as close as possible to, but not larger than the other number 20.

$$
\begin{array}{ccc}
 & 1 & 24 \\
 & 2 & 48 \\
* & 4 & 96 \\
 & 8 & 192 \\
* & 16 & 384
\end{array}
$$

Notice that since 32 > 20, there is no need to go past 16.

> *Step 2* Then subtract the left-side numbers from 20 until you reach 0. Star the left-side numbers that are being subtracted.

$$
\begin{array}{r}
20 \\
-\ \ 16 \\
\hline
4 \\
-\ \ 4 \\
\hline
0
\end{array}
$$

Note: These subtractions show that 20 can be written as the sum of the "doubling numbers" 4 and 16.

> *Step 3* To obtain the answer, add the corresponding right-side numbers of the starred positions.

$$
\begin{array}{r}
384 \\
+\ \ \ 96 \\
\hline
480 \quad \blacksquare
\end{array}
$$

This ingenious method secretly relies on the distributive law — namely that for any three numbers a, b, and c, we have

$$
a \times (b + c) = a \times b + a \times c
$$

or more compactly, using the convention that two symbols with no space between them are multiplied, we have

$$a\left(b+c\right)=ab+ac$$

In other words, we can distribute a multiplier sitting outside parentheses containing a sum. So $17 \times 12 = 17 \times (4+8) = 17 \times 4 + 17 \times 8 = 68 + 136 = 204$. Neat.

Can you use the distributive law to quickly find $45 \times 63 + 45 \times 37$?

Bear in mind that the Egyptian dynasties of 3000 B.C. ushered in the dawn of recorded history, making their "small" mathematical discoveries immensely important and admirable. They started from scratch (like the markings on the wolf bone?) and ended with record-keeping, a calendar, geometry, pyramids, and a working knowledge of arithmetic able to handle very large and very small numbers.

Which brings up an interesting question: How did they handle fractions? No doubt, young Egyptian students of that era dreaded fractions as much as today's school children (and some grown-ups).[2]

While it is true that we study fractions like $\frac{4}{5}$ or "four-fifths" after we have mastered the whole numbers four and five ($\frac{4}{5}$ of a pizza entails cutting the pie into five equal pieces and greedily snatching four of them), there is much evidence to suggest that the important fraction $\frac{1}{2}$ preceded sophisticated whole numbers like 241 and 3762. Consider the special words for $\frac{1}{2}$ in many languages which have no reference to "two," like "half" in English and "moitie" in French, as

---

[2]It is said that five out of every four people have trouble with fractions!

opposed to "one-fourth" or "one-fifth," which are obviously related to four and five.

The Egyptians recognized that fractions begin with the so-called *reciprocals* of whole numbers, like $\frac{1}{3}$ or $\frac{1}{8}$. After all, $\frac{3}{7}$ of something implies that we have a prior understanding of $\frac{1}{7}$. In fact, $\frac{3}{7}$ means three of the latter creature. So, with one or two exceptions, the Egyptians denoted reciprocals of integers by placing an *eye* over them. Thus, one-tenth is written

while a more involved fraction like $\frac{1}{123}$ becomes

Unfortunately for the accountants of their forerunner to the Internal Revenue Service, they refused to consider numerators other than one. They insisted on writing fractions such as $\frac{3}{4}$ as sums of distinct reciprocals, that is, $\frac{3}{4} = \frac{1}{2} + \frac{1}{4}$. This usually requires longer sums. $\frac{7}{8}$, for example, can be written $\frac{4}{8} + \frac{2}{8} + \frac{1}{8}$, or after reducing to lowest terms, $\frac{1}{2} + \frac{1}{4} + \frac{1}{8}$. These fractions are called *unit fractions*.

Example 1-2 *Write $\frac{7}{18}$ in unit fractions (the Egyptian method).*

The method consists of multiplying basic unit fractions $(\frac{1}{2}, \frac{1}{3}, \frac{1}{4}, \frac{1}{5}, \ldots)$ on the denominator to produce numbers that will add up to the numerator.

denominator =	18	
$\frac{1}{2}$	9	(too big)
$\frac{1}{3}$	6	(need 1 more)
$\frac{1}{18}$	1	
	7	

So $\frac{7}{18} = \frac{1}{3} + \frac{1}{18}$. ∎

Would you like to try this for $\frac{123}{500}$? You should finally get $\frac{1}{5} + \frac{1}{25} + \frac{1}{250} + \frac{1}{500}$. This procedure was a computational nightmare and suffered the same ultimate fate as the kingdoms of ancient Egypt. As we will see in the next chapter, the Mesopotamians (like the Babylonians, for example) had a much better idea which, in essence, survives to the present day.

The Egyptian division algorithm indicates that they understood that division is the reverse of multiplication – in much the same way that subtraction is the reverse of addition. The question "What is ten minus three?" can be rephrased "What must be added to three to yield ten?" (The answer is left as an exercise!)

Similarly, the question "What is twenty-four divided by four?" becomes "What must be multiplied by four to yield twenty-four?" – and is handled in the following manner: Put 1 and 4 at the head of two columns, 1 on the left and 4 on the right, and repeatedly double both sides (sound familiar?) until some of the numbers on the right add up to 24. See the work in Figure 1-2.

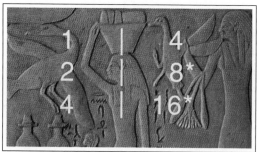

Figure 1-2: Egyptian Division

The asterisks near 8 and 16 indicate that they add up to 24, so we add the numbers opposite them, 2 and 4, to get the answer, 6. The alert reader might notice that many divisions don't work as neatly, like dividing six into twenty-five. This made fractions all the more necessary. In order to see how the Egyptians would have handled it, let's consider a harder example, one with a remainder.

Example 1-3 *Divide 295 by 30 using the Egyptian method of division.*

    Step 1 Start with 1 and the divisor 30. Keep doubling both numbers until the right side gets as close as possible to, but not larger than 295.

$$
\begin{array}{rrl}
1 & 30 & * \\
2 & 60 & \\
4 & 120 & \\
8 & 240 & *
\end{array}
$$

    Step 2 Then subtract the right side numbers from 295 until you can no longer subtract. What is left will be the *remainder*. Star the right-side numbers that are being subtracted.

$$
\begin{array}{rr}
 & 295 \\
- & 240 \\
\hline
 & 55 \\
- & 30 \\
\hline
 & 25
\end{array}
$$

    Step 3 To obtain the answer or *quotient*, add the corresponding left-side numbers of the starred positions.

$$
\begin{array}{rr}
 & 8 \\
+ & 1 \\
\hline
 & 9
\end{array}
$$

Making the answer 9 R 25 or $9\frac{25}{30} = 9\frac{5}{6}$, which the egyptians would have written using unit fractions as $9 + \frac{1}{2} + \frac{1}{3}$. ∎

    One reason the ancient Egyptians had to deal with multiplication involved geometry and measurement. Measurement involves questions like "how much," "how big," "how fast," and "how heavy." The mathematician must then conjure up a "unit" which translates the above questions into "how many *cupfuls*," "how many *inches*," "how many *miles per hour*," and "how many *pounds*." The Egyptians took an enormously giant step by inventing a unit of area from a unit of length by forming a square unit of area! Let's use square feet for simplicity. A foot is a unit of length – but a tile of length and width one foot, that is, a unit square tile, can be said to have area one (one square foot, that is). Now, a rectangular room of length 30 feet and width 20 feet can be tiled with 30 rows of twenty tiles each. Instead of repeatedly adding twenty thirty times, we have that $30 \times 20 = 600$, and the room has a floor area of 600 square feet. "How much area" became "how many square feet" and this is how we measure area today!

As we noted above, the transition from nomadic hunter/gatherer to agrarian city-dweller created a surplus of goods which, in turn, enabled an entire class of people, usually priests, to abstain from production and to concentrate instead on reading, writing, record-keeping, and astronomy (including its early counterpart, astrology).

The night sky was more than just a feast for the eye; it was one of the birthplaces of mathematics! No one knows how long it took, but the Egyptian priests eventually noticed the *periodic* (repetitive) behavior of the *trajectories* (paths) of heavenly bodies. Of course, the earliest humans couldn't help but notice the regular progression of night and day,[3] or the fondness of the sun for routine daily behavior – rising in the east and setting in the west.

Another prominent heavenly body was, no doubt, noticed many "moons" ago! In fact, the moon seemed to inexplicably grow and shrink in a regular pattern discernable to all, with the possible exception of a handful of "lunatics."

---

[3]Wouldn't that be a great title for a song?

The equal time intervals between "new moons," approximately 28 days, afforded the ancient civilizations a means of time measurement that is still the basis of some calendars today (like the Jewish calendar). The Egyptians, like any other intellectual culture, needed a calendar as well, for various reasons, such as

1. knowing when to plant and harvest crops,

2. predicting the annual flooding of the Nile River, and

3. recording important events, like the Pharaoh's birthday.

Perhaps it should be mentioned that the flooding of the Nile was tied to the heliacal rising of Sirius, as observed by Egyptian astronomers, and not the calendar since it did not remain in sync with sun.

Nonetheless, having a calendar required observing the shift on the horizon of the rising of the sun and several prominent stars and planets. The eye sweeps out a huge circle as it beholds the entire horizon. Points on the horizon can, therefore, be measured by the angle between the observer's line of sight and a fixed line. See Figure 1-3.

Figure 1-3

So perhaps geometry should be renamed cosmometry – measuring the cosmos! No doubt, Egyptian geometry was not confined to land surveying and architecture. It played an integral part in locating planets in the sky even after they rose. Imagine, if you will, as the ancients did, that the sky is an enormous hemisphere and Earth is a flat disc sharing a common circular boundary with the sky. An observer looking for the North Star at 1:00 A.M., while taking a break from doing math homework, would use two angles. The first locates a point on the common boundary of the disc and hemisphere, that is, the horizon, and the second is the star's "angle of elevation." See Figure 1-4.

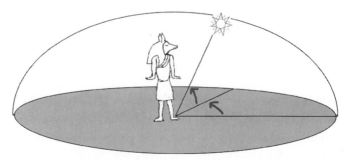

Figure 1-4

The Egyptian year consisted of twelve months of thirty days followed by five feast days with a total of 365 days. Their day was divided into twelve equal parts obtained by determining the position of the sun as it crossed the sky. If we adhered to this convention today, our hours would be longer in the summer than in the winter – as if the heat made them expand.

Sky watching was a religious activity. Virtually all civilizations, then and now, have had some form of deity located in the heavens, from the Egyptians who worshiped the sun god Ra,[4] the Mesopotamians who worshiped Ishtar who lived on Venus, the Greeks and Romans of antiquity (not the name of a country) and their various divine denizens of the heavens, to today's cultures that worship God knows whom. Indeed, early man dreaded the solar eclipse believing it to herald the arrival of a malevolent, hungry deity out to make a noontime snack out of the sun. Bright stars and planets foretold the unfolding of major events like the coming of Christ.

In today's scientific age, observe how many "modern" humans consult their daily horoscopes – decades after the "giant leap for mankind" on the moon.

---

[4]Hence the cheer, "Ra, Ra, Ra!"

Throughout this text, we shall see how math and astronomy have influenced each other and the way we view the universe and our place in it. Cosmological issues, such as "Does Earth revolve around the sun or is it the other way around?" became "burning" issues to the Roman Inquisition while thinkers like Copernicus deftly sidestepped trouble by asserting that God, being an elegant geometer, designed a simple model in which the earth circled the sun – thereby reducing the eighty or so calculations necessary to describe the solar system using the geocentric theory supported by the church to a mere thirty. He also had the good sense to publish his work on the heliocentric theory posthumously. But, we are getting ahead of ourselves. More on that topic later.

1. Eventually, primitive man needed to count. In order to count bigger numbers, he grouped them into groups of five. Explain why ⳾⳾⳾ ⳾⳾⳾ is easier to count than ‖‖‖‖‖‖‖‖‖ .

2. Write the following numbers using hieroglyphics:

   (a) 100
   (b) 55
   (c) 1,024
   (d) 234
   (e) 88
   (f) 7,686

3. Write the following numbers using hieroglyphics:

   (a) 327
   (b) 1,492
   (c) 123,470
   (d) 234,633
   (e) 1,111,111
   (f) 34,209

4. Add the following after rewriting the numbers using hieroglyphics:

   (a) $234 + 765$ (b) $4,555 + 5,648$
   (c) $36,486 + 9,018$ (d) $546,386 + 675,201$
   (e) $258,586 + 987,475$ (f) $465,999 + 555,486$

5. The "doubling" method of Egyptian multiplication requires
   writing any whole number as the sum of powers of two, with
   no repetitions. For example, $45 = 32 + 8 + 4 + 1$. Try this for
   each of the following. [Recall, the powers of two are 1, 2, 4, 8,
   16, 32, 64, 128, ...]

   (a) 73   (b) 52
   (c) 98   (d) 151

   (e) How do you know this can be done for any number?

6. Multiply using the Egyptian method of doubling:

   (a) 19×29 (b) 25×73 (c) 13×21 (d) 71×211

7. Multiply using the Egyptian method of doubling:

   (a) 23×25 (b) 34×82 (c) 31×36 (d) 55×107

8. Write the following fractions using hieroglyphics:

   (a) $\frac{1}{3}$ (b) $\frac{1}{5}$ (c) $\frac{1}{63}$ (d) $\frac{1}{1,124}$

9. Write these fractions as sums of fractions with unit numerators
   using the Egyptian method:

   (a) $\frac{3}{10}$ (b) $\frac{4}{5}$ (c) $\frac{111}{200}$ (d) $\frac{7}{50}$

10. Write these fractions as sums of fractions with unit numerators:

    (a) $\frac{13}{40}$ (b) $\frac{9}{46}$ (c) $\frac{4}{21}$ (d) $\frac{4}{11}$

11. Sometimes it is possible to use the proper divisors of the de-
    nominator that add up to the numerator to split a fraction
    into unit fractions. Take $\frac{8}{15}$. The proper divisors of 15 are 1,
    3, and 5. The 3 and 5 add up to 8, giving us the fact that
    $8/15 = 5/15 + 3/15 = 1/3 + 1/5$. Use this technique to write
    the following fractions as sums of unit fractions:

    (a) $\frac{3}{10}$ (b) $\frac{4}{5}$ (c) $\frac{111}{200}$ (d) $\frac{7}{50}$

12. Show that for any positive integer (whole number) n,

$$\frac{1}{n} = \frac{1}{n+1} + \frac{1}{n(n+1)}$$

For example, if we let $n = 2$, we get $\frac{1}{2} = \frac{1}{3} + \frac{1}{6}$.

13. Using the formula from Exercise 12, write the following "unit" fractions as the sum of two other unit fractions:

(a) $\frac{1}{5}$ (b) $\frac{1}{11}$ (c) $\frac{1}{10}$ (d) $\frac{1}{101}$

14. By repeatedly using the formula of Exercise 12, write the following as sums of three unit fractions. Could one keep increasing the number of unit fractions in these sums?

(a) $\frac{1}{3}$ (b) $\frac{1}{9}$ (c) $\frac{1}{12}$ (d) $\frac{1}{20}$

15. Suppose an alien civilization on a distant planet (in a distant solar system, of course) uses the following base-five notation. (They have two hands with two and a half fingers each.)

1	%
5	#
25	&
125	$
625	@

(a) Compare and contrast their system with that of ancient Egypt. Why would a base-five system need symbols for 25, 125, and 625?

(b) Write 18, 47, 283, 1005, and 3000 in "alienese."

(c) What is the largest number that could be written, assuming that no symbol appears more than four times?

16. Perform the following divisions using the Egyptian method:

(a) $75 \div 15$ (b) $156 \div 13$ (c) $80 \div 8$ (d) $91 \div 7$

17. Perform the following divisions using the Egyptian method:

(a) $6,598 \div 37$ (b) $805 \div 35$ (c) $528 \div 22$ (d) $1,134 \div 42$

18. From the law $\dfrac{a}{b} \times \dfrac{c}{d} = \dfrac{ac}{bd}$ and the observation that any integer n can be written $\dfrac{n}{1}$, it follows that $\dfrac{1}{3} \times 6 = \dfrac{1}{3} \times \dfrac{6}{1} = \dfrac{6}{3}$. On the other hand, these same rules tell us that $\dfrac{3}{3} \times \dfrac{2}{1} = \dfrac{6}{3}$. Now $\dfrac{3}{3} = 1$, right? So this last equation can be written $2 = \dfrac{6}{3}$. This is one of the ways of justifying the operation with fractions known as *reducing to lowest terms*. One more example. To reduce $\dfrac{20}{5}$, we have $\dfrac{20}{5} = \dfrac{5}{5} \times \dfrac{4}{1} = 4$. This procedure can be summarized abstractly by the equation $\dfrac{ab}{ac} = \dfrac{b}{c}$. On the other hand, can't a fraction be interpreted as a division? Thus, $\dfrac{6}{3}$ is the same as $6 \div 3$, which is, of course, 2.

   (a) Explain how $6 \div 3$ is the same, arithmetically and conceptually, as taking $\frac{1}{3}$ of 6. How would you interpret the same problem written $6 \times \frac{1}{3}$? This is clearly a multiplication problem, not a division problem.

   (b) In general, state three different interpretations of the expression $\frac{a}{b}$ and explain why they all have the same answer. Illustrate your work using the equation $\frac{15}{5} = 3$.

   (c) Rewrite this previous equation as $\frac{15}{5} = \frac{3}{1}$ and find a fourth interpretation by viewing this last equation as the equivalence of the ratios: 15 is to 5 as 3 is to 1.

19. The *Egyptian method of False Position* is a guess method for solving linear equations. For example, a number plus its double plus its third, yields 20. What is the number? If we guess 12, and plug this number into the first part of the problem we obtain $12 + 2(12) + \dfrac{12}{3} = 40$. This is too much since we require 20. But if we form the "magic number" $\dfrac{20}{40}$ and multiply this by our original guess 12, we get the actual answer 6. Use the method of False Position to solve the following: A number plus one-seventh the number is 24. Find the number.

20. Using the Egyptian method of False Position solve each of the following:

   (a) $3x + \dfrac{x}{4} = 65$

   (b) $3x - \dfrac{x}{4} + \dfrac{3}{8}x = 10$

   (c) $\dfrac{x}{2} + \dfrac{x}{4} + \dfrac{x}{8} + \dfrac{x}{16} = 75$

21. A *ciphered number system* is one in which many different symbols are used to represent the digits 1, 2, 3, ... , 9 as well as 10, 20, 30, ... , 90, and 100, 200, 300, ... , 900, and so on. The *Egyptian hieratic system* was one such system. Investigate this number system on the Internet and write a short paper summarizing your results.

Suggestions for Further Reading

1. Aaboe, Asger. *Episodes from the Early History of Mathematics*. Random House, New York, 1964.

2. Cajori, Florian. *A History of Mathematics*. Macmillan, New York, 1919.

3. Gillings, Richard J. *Mathematics in the Time of the Pharaohs*. Dover, New York, 1982.

4. Menninger, Karl. *Number Words and Number Symbols; A Cultural History of Numbers*. Dover, New York, 1992.

5. Neugebauer, Otto. *The Exact Sciences in Antiquity*. Dover, New York, 1969.

6. Ore, Oystein. *Number Theory and Its History*. Dover, New York, 1988.

Chapter 2

Mesopotamia Here We Come

The region between the Tigris and Euphrates Rivers is known as the "Fertile Crescent" (modern southern Iraq from around Baghdad to the Persian Gulf), which, like the Nile, is termed a "cradle of civilization." Approximately five thousand years ago, the Sumerians developed a system of pictographic writing similar to Egyptian hieroglyphics, including a written number system. Sometime around 2300 B.C., a Semitic tribe known as the Akkadians invaded the Fertile Crescent and replaced the Sumerians, while retaining much of their writing, which by then had evolved into a sophisticated system involving only two symbols — a horizontal wedge and a vertical one. These symbols were impressed onto clay tablets with a stylus, after which the tablets were baked. Fortunately, an incredible number of these tablets have survived (clay beats papyrus in the game "clay, scissors, papyrus") and scholars can read them. In 1870, the Behistun Cliffs were found to contain a description of a victory of the Persian king Cyrus written in Babylonian (similar to Akkadian) and Persian, which was already understood. After years of study, the *"cuneiform"* wedge marks of the Akkadians were finally understood.

The Mesopotamian culture is often called Babylonian, after the large metropolis of that name. We could "babble on"[1] and on about their many fine achievements in architecture, irrigation, and commerce, but it is their mathematics that is truly remarkable, dwarfing that of other contemporary civilizations. One might not be impressed by their use of a vertical mark for "one" and a horizontal mark for "ten" – ten being a common unit in the mathematics of many societies, including Egypt, China, Rome, and our own society today. On the other hand, they were the first to employ a "positional" system which, except for minor changes, survives to this day!

Let's remind ourselves how our current number system works. It does not suffice to say that it is based on grouping by tens. The

---

[1] The authors would like to apologize for the easy pun, but we couldn't resist.

Egyptians did this – yet we have left them in the dust by taking a giant step forward to the "position system." We require only ten symbols: 0, 1, 2, 3, 4, 5, 6, 7, 8, and 9. Nevertheless, we can handle numbers of any size without the need to define a new symbol. This is because the value of a number is determined not just by the symbol. We must note the *position* of the symbol as well. The two 3's in the number 373 represent different quantities. You would rather have three hundred dollars than three dollars, right? To summarize, our number system employs a mere ten symbols, whose values depend on their position in the number. Moving one digit to the left multiplies its place value by ten, while moving to the right (not surprisingly) divides its place value by ten.

Observe, by the way, that this is true on both sides of the decimal point! In the number 3.1416, the 1 near the 6 is worth only one hundredth of the 1 near the 3. There is no number in the entire universe that is too large or too small for our clever (ten-digit!) number system (of Hindu-Arabic origin, by the way). We call our system the *decimal* system, because ten is the base.

The Babylonians used instead the *sexagesimal* system because they chose 60 as their base. While we are not sure why, we are fairly certain they did not have 60 fingers. One theory (which is very popular) is that 60 has a multitude of factors, that is, many numbers go into 60. Put another way, $60 can be divided without coin among 2, 3, 4, 5, 6, 10, 12, 15, 20, or 30 people. We shall follow the common practice of using commas to separate groups. Thus $(3, 50)_{60}$ shall mean 3 sixties and 50 ones for a total of 230. What does $(2, 3, 50)_{60}$ mean? Well in our position system, 357 means 3 hundreds, 5 tens, and 7 ones, right? Each column is ten times more valuable than its neighbor. In the same way, each column to the left in the Babylonian system is sixty times bigger! In the number $(2, 3, 50)_{60}$, the 2 represents 2 3600's – because $60 \times 60 = 3600$. The next column to the left would represent 60×3600 or 216000.

The Babylonians only used two symbols: a vertical mark for 1 and a horizontal mark for 10. Thus, the number 230, which we denoted by $(3, 50)_{60}$ would look like this:

Let's take a look at some examples.

Example 2-1 *Write 28 in Babylonian cuneiform.*

Again, the Babylonians used wedges and wrote in base-60. They used for 1 and for 10. They never wrote more than three of a particular symbol in a row or column. Thus, 28 can be written as

They sometimes used

to represent subtraction, so 28 which is 30 − 2 could be written as

This subtraction symbol was not used to represent numbers in the fifties because that would have required sixty which is not used in base-sixty. ∎

Example 2-2 *Translate the following from Babylonian cuneiform into our number system.*

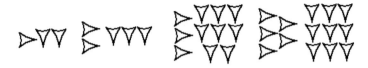

In other words, convert $(12, 23, 38, 59)_{60}$ *to base-10.*

Remember that the first number in the parentheses on the right, in this case the 59, is in the "ones" place and each place to the left has a place value 60 times greater. This makes the 12's place worth $60^3 = 216{,}000$. The number is calculated as follows:

$$
\begin{array}{rll}
12 \times 60^3 = & 12 \times 216{,}000 = & 2{,}592{,}000 \\
23 \times 60^2 = & 23 \times 3600 = & 82{,}800 \\
38 \times 60^1 = & 38 \times 60 = & 2{,}280 \\
+ \quad 59 \times 60^0 = & 59 \times 1 = & 59 \\
\hline
& & 2{,}677{,}139 \quad \blacksquare
\end{array}
$$

Example 2-3 *Write* $7{,}820$ *in base-60.*

Method 1: The largest power of 60 that goes into 7820 is 3600. It goes in twice, leaving a remainder of $7820 - 2 \times 3600 = 7820 - 7200 = 620$. Now 60 goes into 620 ten times leaving a remainder of $620 - 10 \times 60 = 620 - 600 = 20$. Then this goes into the units place. The final answer is $(2, 10, 20)_{60}$.

Method 2: The answer can also be obtained through repeated division by the base-60 and noting the remainders. Now, $7{,}820 \div 60 = 130 \ R \ 20$, $130 \div 60 = 2 \ R \ 10$, and $2 \div 60 = 0 \ R \ 2$. The answer is obtained by taking the remainders in reverse order. $(2, 10, 20)_{60}$. $\blacksquare$

Addition and subtraction were done by working in columns and often required "borrowing" (just as the Egyptians required). Once they figured out their position system, they took another giant step. They treated fractions as a continuation of the base-60 system! A 1 in the column to the right of the units column has the value $\frac{1}{60}$, while a 1 in the next column has the value $\frac{1}{3600}$, and so on. Babylonian fractions were so accurate and easy to use that the Greeks, whose number system for whole numbers was markedly different, employed them a thousand years later.

Unfortunately, the Babylonians neglected to do two things:

1. They had no symbol for zero, no placeholder to indicate that a column had no entry. This is similar to my writing 301 as 3 1. After all since there are no "tens" in 301, why put the zero there? The obvious answer is that without the zero, we will conclude that the "3" represents tens. The Babylonians had to use context and guesswork to read their numbers correctly!

2. Another "small" oversight: They had no decimal point (or sexagesimal point?). Imagine leaving out the decimal point in the price of a \$17.95 steak and scaring away all your customers. The decimal point lets us know where the fractions begin! This problem, too, required caution in the way the Babylonians read their numbers.

We shall use a semicolon to separate the whole number columns from the fractional ones (like the decimal point of today's base-ten arithmetic).

Fraction	Sexagesimal	Fraction	Sexagesimal
$\frac{1}{2} = \frac{30}{60}$	0; 30	$\frac{1}{18} = \frac{3}{60} + \frac{20}{3600}$	0; 3, 20
$\frac{1}{3} = \frac{20}{60}$	0; 20	$\frac{1}{20} = \frac{3}{60}$	0; 3
$\frac{1}{4} = \frac{15}{60}$	0; 15	$\frac{1}{24} = \frac{2}{60} + \frac{30}{3600}$	0; 2, 30
$\frac{1}{5} = \frac{12}{60}$	0; 12	$\frac{1}{25} = \frac{2}{60} + \frac{24}{3600}$	0; 2, 24
$\frac{1}{6} = \frac{10}{60}$	0; 10	$\frac{1}{30} = \frac{2}{60}$	0; 2
$\frac{1}{8} = \frac{7}{60} + \frac{30}{3600}$	0; 7, 30	$\frac{1}{32} = \frac{1}{60} + \frac{52}{3600} + \frac{30}{216000}$	0; 1, 52, 30
$\frac{1}{9} = \frac{6}{60} + \frac{40}{3600}$	0; 6, 40	$\frac{1}{36} = \frac{1}{60} + \frac{40}{3600}$	0; 1, 40
$\frac{1}{10} = \frac{6}{60}$	0; 6	$\frac{1}{40} = \frac{1}{60} + \frac{30}{3600}$	0; 1, 30
$\frac{1}{12} = \frac{5}{60}$	0; 5	$\frac{1}{45} = \frac{1}{60} + \frac{20}{3600}$	0; 1, 20
$\frac{1}{15} = \frac{4}{60}$	0; 4	$\frac{1}{48} = \frac{1}{60} + \frac{15}{3600}$	0; 1, 15
$\frac{1}{16} = \frac{3}{60} + \frac{45}{3600}$	0; 3, 45	$\frac{1}{50} = \frac{1}{60} + \frac{12}{3600}$	0; 1, 12

Table I: Babylonian Arithmetic

NOTE: Addition, subtraction, and multiplication are performed as in our system but remember the base is 60, so you carry and borrow based on 60. Multiplying the dividend by the reciprocal of the divisor performs division. To simplify the division process, tables of reciprocals were used for those numbers that have a terminating sexagesimal fraction. The Babylonians tried to avoid using fractions with nonterminating sexagesimal fractions, like 1/7 and 1/11, and so forth. The semicolon (;) is used to separate whole numbers from the fractional part.

The Babylonians multiplied much the way we do, including the use of "carrying," but their division was noteworthy. They used a table of reciprocals. (See Table I.) Recall that the reciprocal of n is $\frac{1}{n}$, which has the effect of turning it upside-down. The reciprocal of $\frac{2}{5}$ is $\frac{5}{2}$, for example. Now, since 2 can be written $\frac{2}{1}$, its reciprocal is $\frac{1}{2}$. The reciprocal of 4 is $\frac{1}{4}$. Why the interest in reciprocals? Because division by a number n is the same as multiplication by its reciprocal $\frac{1}{n}$. To divide by four, one can multiply by one-fourth! Herein lies one of the strengths of the base-sixty system. Many fractional parts of 60 are whole numbers. One half of 60 is 30. So $\frac{1}{2} = \frac{30}{60}$ or $(0; 30)_{60}$ in the way we have been writing Babylonian numbers. One-third of 60 is 20, so $\frac{1}{3} = \frac{20}{60}$ or $(0; 20)_{60}$.

Example 2-4 *Divide 40/12 using the Babylonian method of division.*

The Babylonians would have written the division as a multiplication by the reciprocal. So $\frac{40}{12} = \frac{1}{12} \times 40 = (0; 5)_{60} \times 40$. Now $5 \times 40 = 200$ which when written in base-60 is $(3, 20)_{60}$. This means that we carry the 3 to the next position giving us as an answer $(3; 20)_{60}$. ■

Example 2-5 *Divide 100/8 using the Babylonian method of division.*

Again, rewriting the division as a multiplication by the reciprocal gives $\frac{100}{8} = \frac{1}{8} \times 100 = (0; 7, 30)_{60} \times 100$. Multiplying $30 \times 100 = 3000$, which expressed in base-60 is $(50, 0)_{60}$. This means that we would carry the 50 to the next position. Since there is a 7 there, we would multiply $7 \times 100 + 50 = 750$ which can be written as $(12, 30)_{60}$ in base-60 giving the final answer $(12; 30, 0)_{60}$. ■

The Babylonians divided their daylight into twelve hours as the Egyptians did. Their fixation on sixty manifested itself in their dividing the hour into 60 minutes ("minute" for "small") and the minute into 60 seconds (the "second" division of the hour). So our clocks are Babylonian and Egyptian. It is strange that this practice persists in the decimal age! (Perhaps it isn't so strange – Americans still resist the "metric" system, preferring 12 inches in a foot, 3 feet in a yard, etc.)

The Babylonians developed their commercial mathematics and computed interest tables for financing loans! They extracted square and cube roots and tackled simple "word problems" requiring the solution of quadratic equations. While their geometry sufficed, they were particularly skilled in algebra and computing with fractions. Their algebra, however, lacked the notation that facilitates ours, that is, "let x be Johnny's age, and $x + 4$ be Sally's age. Then in two years, the sum of their ages will be $(x + 2) + (x + 6) = 2x + 8$." Rather, theirs was a procedural algebra. Thus, if twice a quantity of apples augmented by ten apples yields fifty, then a Babylonian might tell you to "taketh away ten apples from the fifty and then ye must take half the remaining quantity. Then by the gods, thou shalt be left with the original amount."

Of course this is exactly what we do today:

$$2x + 10 = 50$$
$$-10 = -10$$
$$2x = 40$$
$$\left(\frac{1}{2}\right) 2x = \left(\frac{1}{2}\right) 40$$
$$x = 20$$

Babylonian cosmology was a mixture of science and religion. It was believed that the gods resided on the planets, from which it followed that their location would influence the actions of the gods in human affairs – hence, the birth of astrology. The Greek and Roman beliefs in this regard are Babylonian in origin. The Babylonians assumed that Earth was the center of the universe and everything revolved around it. This cosmology was later adopted by the early Christian thinkers and perpetuated for a thousand years after the fall of Rome. The "geocentric theory" was consistent with the Christian belief that the Son of God was born at the center of the universe.

Among the many tablets dug up in the Fertile Crescent, one of them, called *Plimpton 322*, contained many pairs of numbers with a curious property. The square root of the difference of their squares is a whole number. You might recall the famous 3, 4, 5 triple that has this property, that is, $3^2 + 4^2 = 5^2$. We will meet such triples when we study the Greek mathematician, philosopher, and cult figure Pythagoras and his famous theorem which says that the sum of the squares of the two shorter sides of a right triangle equals the square of the hypotenuse, or as it is usually presented,

$$a^2 + b^2 = c^2$$

where a and b represent the lengths of the shorter sides (called the *legs*) and c is the length of the hypotenuse.

Now things usually get messy when the two shorter sides (or "legs") are randomly selected. If they are 2 and 5, for example, then the sum of their squares is 29 (i.e., $4 + 25 = 29$), implying that the hypotenuse is $\sqrt{29}$. Without a calculator, this poses a significant challenge, though it's easy to say it is between 5 and 6. You might also recall the triple 5, 12, 13 – an amazing triple. While these triples might result from a lucky guess or two, the triples on the recovered clay tablet go into the thousands, for example 2291, 2700, and 3541, indicating that the Babylonians had a method probably similar to that of the Greeks.

We close this chapter with an amazing tale of cleverness. The Babylonians managed to compute square roots quite accurately, as is exemplified by their calculating $\sqrt{2}$ to seven places of accuracy. We surmise that they used a method of averaging that goes something like this. Suppose we want $\sqrt{363}$. Let us begin with a guess of 15. This is obviously too small since $15 \times 15 = 225$, whereas the square root of a number times itself should yield the original number, in this case 363. Furthermore, if we calculate $363 \div 15$, we get 24.2 (which is too big to be $\sqrt{363}$). Well here is a brilliant idea. Since 15 is too small and 24.2 is too big, let's take their average, that is, half of their sum, or $(24.2 + 15)/2$, which is 19.6. This is much closer to the answer than 15 or 24.2. Let's repeat this procedure one more time. Since $363/19.6 = 18.520408163\ldots \cong 18.52$, and once again 18.52 is too small and 19.6 is too big, let's average them: $(18.52 + 19.6)/2 = 19.06$, which is even closer to the "true" answer of $19.05255888325\ldots$. This *algorithm* (procedure) can be repeated as often as one wants, yielding a sequence of increasingly accurate answers.

Another way of calculating square roots is to use the formula

$$\sqrt{N} = S + \frac{E}{2S} - \frac{E^2}{8S^3}$$

where $N = S^2 + E$, that is, S is the largest number whose square is less than N and E is the difference between this perfect square and N. While the Babylonians didn't have a formula, don't forget they had a procedural algebra that could describe this formula. Let's see how to use it in the next example.

Example 2-6 *Compute $\sqrt{363}$ using the "Babylonian" formula.*

Since $363 = 19^2 + 2$, we have that $S = 19$ and $E = 2$. Plugging these into the formula gives $\sqrt{363} = 19 + \frac{2}{38} - \frac{4}{54{,}872} = 19.0525586820236\ldots$ which is accurate to 6 decimal places. ∎

In summary, Babylonian mathematics was extremely sophisticated for its time. Our position system today is virtually identical to theirs, as is our system of time measurement. The next great ancient civilization we study, Classical Greece, inherited Babylonian algebra and Egyptian geometry, no doubt, through the travels of her scholars and merchants. It should be noted that very little is known about the mathematics of ancient China. In 213 B.C., the emperor Shi Huang of the Chin dynasty had all of the manuscripts of the kingdom burned. Fortunately, a copy of a work titled *Arithmetic in Nine Sections* survived. Written before 1000 B.C., it contains mathematics roughly on par with that of Egypt and Babylon.

1. Write the following numbers in Babylonian cuneiform using $\bigvee$ for 1 and $\triangleright$ for 10:

   (a) 26   (b) 46   (c) 32   (d) 17   (e) 11   (f) 58

2. Write the following numbers using the Babylonian position system. Use our numerals separated by commas and use a semi-colon, if necessary, to indicate where the fractional part begins.

   (a) 90   (b) 75   (c) 3660
   (d) 7200   (e) 7325   (f) $\frac{3}{4}$

3. Write the following numbers using the Babylonian position system. Use our numerals separated by commas and use a semi-colon, if necessary, to indicate where the fractional part begins.

   (a) 47   (b) 78   (c) 3662   (d) $\frac{2}{3}$ (e) $\frac{11}{4}$ (f) $\frac{21}{15}$

4. Express these Babylonian numbers using our modern number system:

   (a) 1, 1   (b) 1, 1, 0; 10   (c) 1; 30
   (d) 1, 2, 3; 30, 20   (e) 1, 0; 0, 1

5. Express these Babylonian numbers using our modern number system:

   (a) 1, 3   (b) 3, 1, 2; 10   (c) 1, 1, 1; 6, 36
   (d) 4, 3; 0, 0, 1   (e) 0; 30, 23

6. Find the hypotenuse of a triangle whose legs are

   (a) 7 and 24   (b) 9 and 40

7. (a) To multiply a decimal number by ten, we shift the decimal point one unit to the right. Thus, $10 \times 3.1416 = 31.416$. See if you can devise a rule for multiplying a Babylonian number by 60. (b) How would you divide a Babylonian number by 60?

8. Calculate $\sqrt{200}$ in two ways. Firstly, use the Babylonian method of averaging until your answer agrees with that of your pocket calculator. How many times did you perform the averaging? Secondly, use the Babylonian formula of Example 2-6.

9. Change the following fractions into fractions with denominator 60. Then write them in the Babylonian style. (Do not consult the table of reciprocals.)

   (a) $\frac{1}{5}$ (b) $\frac{1}{6}$ (c) $\frac{1}{10}$ (d) $\frac{1}{12}$
   (e) $\frac{1}{15}$ (f) $\frac{1}{20}$ (g) $\frac{1}{30}$ (h) $\frac{1}{60}$

10. Divide each of the following using the Babylonian method of division:

    (a) $13 \div 5$ (b) $19 \div 10$ (c) $7 \div 24$ (d) $23 \div 12$

11. Divide each of the following using the Babylonian method of division:

    (a) $29 \div 16$ (b) $80 \div 15$ (c) $5 \div 24$ (d) $17 \div 12$

12. Calculate the following square roots using the Babylonian method of averaging. Find the third approximation.

    (a) $\sqrt{8}$ (b) $\sqrt{15}$ (c) $\sqrt{37}$ (d) $\sqrt{145}$

13. Calculate the following square roots using the Babylonian formula of Example 2-6:

    (a) $\sqrt{8}$ (b) $\sqrt{15}$ (c) $\sqrt{37}$ (d) $\sqrt{145}$

14. If m is any natural number greater than 1, calculate $\frac{1}{m} + \frac{1}{m+2} = \frac{a}{b}$ where a and b are reduced. This will yield a Pythagorean triple with $c = \sqrt{a^2 + b^2}$. Find the triples for $m = 2, 3, 4, 5$, and 6.

15. Use the formula of Exercise 14 to find the Pythagorean triples for $m = 7, 8, 9$, and 10.

16. The Babylonians had a method of solving quadratic equations similar to that of *completing the square* that you may be familiar with from an algebra course you may have taken. Consult a college algebra textbook and write a one-page report on the steps to solve a quadratic equation by completing the square. Be sure to include examples.

17. Write a short paper on the *Plimpton 322* tablet.

18. This chapter presented two methods in Example 2-3 for converting a base-10 number into base-60. Explain why both methods yield the same result.

Suggestions for Further Reading

1. Anglin, W. S. *Mathematics: A Concise History and Philosophy.* Springer-Verlag, New York, 1994.

2. Ball, W. W. Rouse. *A Short Account of the History of Mathematics.* Dover, New York, 1960.

3. Bell, E. T. *The Development of Mathematics.* Dover, New York, 1992.

4. Burton, David M. *The History of Mathematics: An Introduction.* McGraw-Hill, New York, 1998.

5. Eves, Howard Whitely. *An Introduction to the History of Mathematics.* Holt, Rinehart and Winston, New York, 1969.

6. Ifrah, Georges. *The Universal History of Numbers: From Prehistory to the Invention of the Computer.* John Wiley & Sons, New York, 1999.

Chapter 3

Those Incredible Greeks!

In the seventh century B.C., Greece consisted of a collection of independent city-states covering a large area including modern day Greece, Turkey, and a multitude of Mediterranean islands. (This period is called Hellenic, to differentiate it from the later Hellenistic period of the empire resulting from the conquests of Alexander the Great.)

Greek merchant ships sailed the seas, which brought them into contact with the civilizations of Egypt, Phoenicia, and Babylon, to name just a few. This brought Magna Greece ("Greater Greece") prosperity and a steady influx of cultural influences like Egyptian geometry and Babylonian algebra and commercial arithmetic. Moreover, prosperous Greek society accumulated enough wealth to support a leisure class, intellectuals, and artists with enough time on their hands to study mathematics for its own sake! It may come as a surprise that modern mathematics was born in that setting.

This raises a wonderful and timely question: What is *modern* mathematics? While the answer will require most of this book, let us say here that two major characteristics are obvious to any high school graduate:

1. Mathematical "truths" must be proven! A theorem is not a theorem until someone supplies a proof. Before that, it is merely a conjecture, a hypothesis, or a supposition.

2. Mathematics builds on itself. It has a structure. One begins with definitions, axiomatic truths, and basic assumptions and then moves on to consequences or theorems, which, in turn, are used to prove more theorems (often more advanced). An algebraic truth might be utilized to prove a geometric fact. A technique for solving an equation might be employed to find the x-intercept of a straight line whose slope and y-intercept are known. This continuity in mathematics often upsets students who in a college calculus course must recall trigonometric facts they learned in high school (a cruel twist of fate!).

These properties of modern mathematics are a small part of the rich legacy of Ancient Greece. The man who set the ball rolling was a philosopher named Thales[1] who flourished around 600 B.C. Although very little is known for certain about Thales, we can say he was the first to introduce the idea of skepticism and criticism into Greek philosophy, and it is this notion that separates the Greek thinkers from those of earlier civilizations. His philosophy has often been called monism – the belief that everything is one. Many pre-Socratic philosophers were monists, though they differed wildly about the nature of the one thing the entire universe consisted of. Thales observed that water could exist as ice and steam, as well as in a liquid state, leading him to the rather odd hypothesis that the stuff of the universe is water.[2] Before you dismiss Thales as a lunatic, please remember that the oneness of the universe is a very popular idea in the philosophies of the orient to this very day.

You must have heard of the guru who, upon arriving in Manhattan for the first time, points to a hot dog and says to the vendor, "Make me one with everything." He hands the vendor a $20 bill and after waiting a minute inquires, "What about change?" The vendor, also a student of philosophy, replies, "Change must come from within."

Moving right along, Thales was also a mathematician of note.

---

[1]When Thales was asked what is most difficult, he said, "To know thyself." And on being asked what is most easy, he replied, "To give advice."

[2]You might call Thales a hydromonist!

He asserted that mathematical truths must be proven. They must be shown to follow from earlier truths. Among his assertions was the theorem taught today, as "vertical angles are equal." Observe the angles in Figure 3-1 labeled a, b, and c. These letters aren't names of the angles – they are their values in degrees. Now, clearly $a + b = 180°$ and $b + c = 180°$, since $180°$ (reminder: $°$ means degrees) is half of a complete rotation, and a straight line with a point on it chosen as the vertex of the angle can be thought of as the result of a "180-degree turn." A simple transposition of the two equations yields $a = 180° - b$ and $c = 180° - b$, forcing us to conclude that $a = c$. Amazing! Do you realize that no mention was made of the magnitude of the angles? We have a right to assign any values to a, b, and c, as long as $a + b$ and $b + c$ are $180°$.

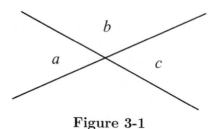

Figure 3-1

We cannot overemphasize the giant step that Thales took with this proof. He demonstrated the all-abiding truth of a general statement that encompasses infinitely many different situations. Two roads crossing in a yellow wood (if they are straight roads) yield a specific instance of this fact, as does a pair of crossed chopsticks lying on a table. The printed symbol X is a silent tribute to the equality of vertical angles! Perhaps the crossing of swords before a fencing match is a salute to Greek geometry. Point noted.

The fact itself was, no doubt, known to Egyptian and Babylonian geometers – but the Greeks *proved* it, and by doing so, advanced mathematics beyond the empirical stage to the theoretical. Geometers drew forth new truths from this one. They observed that a line crossing two given parallel lines, called a *transversal*, makes equal angles with them. Consider the angles labeled a and b in Figure 3-2. They are easily seen to be equal by moving one of the parallel lines slowly toward the other and keeping it parallel to the other given line until the point of intersection coincides with the latter's and the angles come into coincidence. Since we already know that vertical angles are equal, we have that $b = c$ and finally conclude that $a = c$. These angles are usually called "alternate interior angles." We have, therefore, a proof that *alternate interior angles are equal*, which employs in its proof the prior fact (or theorem) that *vertical angles are equal*.

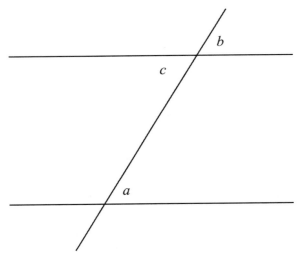

Figure 3-2

Have we made any other assumptions? Absolutely! In both proofs, we have assumed that the two equations $a = b$ and $b = c$ allow us to conclude that $a = c$. Mathematicians call this the *transitive law*. Euclid of Alexandria explicitly stated this some 300 years after Thales. More on this later, when we toast this great geometer, number theorist, and writer, who established the foundations of Greek geometry.[3]

Since this last proof smacks of both geometry and algebra, one sees the influence of both Egyptian and Babylonian mathematics, that is, the geometry of the former and algebra of the latter.

Before proceeding to his next theorem, if you have trouble remembering which pairs of angles in Figure 3-2 (there are eight angles in this figure!) are alternate interior angles, regard the letter Z and try to find it in Figure 3-3. Its angles form a pair of alternating interior angles a and d. The angles of the first-grader backwards Z in the same figure yields the other pair b and c.

---

[3] We shall raise our glasses in toast and exclaim, "Here's lookin' at Euclid!"

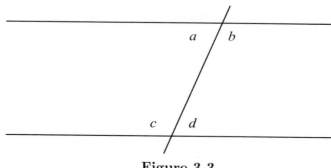

Figure 3-3

It is believed that Thales is also responsible for the incredibly clever and astonishing theorem that the sum of the angles of any triangle is 180°. Consider the triangle of Figure 3-4, whose angles are labeled a, b, and c. We have drawn a line through the upper vertex parallel to the base. This diagram contains two pairs of alternate interior angles, justifying our labeling of the two angles outside of the triangle at the upper vertex a and b. It then follows that $a + b + c = 180°$, since the outer rays of these angles form a straight line.

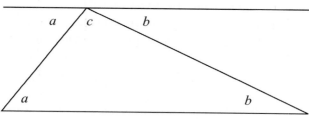

Figure 3-4

Lastly, Thales is alleged to have shown that an angle "inscribed" in a semicircle must be a right angle. The angle $\angle ACB$ in Figure 3-5 demonstrates what we mean by "inscribed in a semicircle." The center of the semicircle is denoted by O and we have drawn line $\overline{OC}$ to assist in the proof. Now Thales also proved that if two sides of a triangle are equal, that is, if it is an *isosceles triangle*, then the base angles (the angles opposite them) are also equal. We shall deal with this when we discuss congruent triangles.

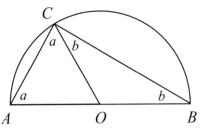

Figure 3-5

Notice that line segments $\overline{OA}$, $\overline{OB}$, and $\overline{OC}$ in Figure 3-5 are equal since they are radii of the circle. Then triangles $\triangle AOC$ and $\triangle BOC$ are isosceles, which in turn allows us to label the two angles, which together comprise $\angle ACB$, a and b. Now we use the fact that the angle sum of triangle ABC is $180°$ to get the equation $a + (a + b) + b = 180°$. The parentheses in this equation remind us that the middle angle ($\angle ACB$) equals $a + b$. We can remove them, thereby changing the equation to $a + a + b + b = 180°$. This becomes $2a + 2b = 180°$, or $2(a + b) = 180°$. Upon dividing both sides of this last equation by 2, we finally get $a + b = 90°$. This, of course, implies that $\angle ACB = 90°$, which is what Thales wanted to show. If you had trouble following this argument, be brave and try it again.[4]

The twenty-first century mathematicians of today's fast-paced world constantly do what Thales did. We develop and then *prove* theorems.

The next great hero of Greek mathematics is Pythagoras,[5] who has been immortalized by the theorem named after him. He came from the island of Samos and probably studied under Thales. An approximate date for his theorem is 540 B.C. He lived during the time of Buddha in India, Lao-Tse in China, and Zoroaster in Persia.

Historical dates, by the way, serve several purposes. Firstly, they put important events in chronological order – an especially important thing in the history of mathematics in which an idea depends on prior ideas. Secondly, they permit us to observe contemporaneous events in two countries or cultures. Thirdly, they make authors seem more scholarly. For most purposes, it suffices to have an approximate

---

[4]If at first you don't succeed, perhaps skydiving is not for you, but mathematics is.

[5]Pythagoras believed in the transmigration of the soul at death into another body, animal or human. So the next time you see someone mistreating an animal, remember that it might be Pythagoras himself.

notion of dates and that is the way they are usually used in this book.

Pythagoras and his followers settled down in Croton – a Greek colony on the Italian peninsula. The Pythagorean brotherhood was a cult movement and their symbol was the pentagram of Figure 3-6, which, by the way, is rich in geometric relationships.

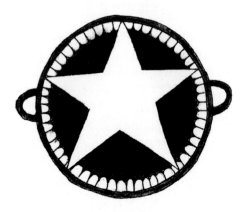

Figure 3-6

Perhaps emulating his teacher, Pythagoras was a philosopher as well as a mathematician, as were many of the early mathematicians. In the monist tradition of the pre-Socratic period, he taught that "everything is number!" This was not a precursor of the frequencies, wavelengths, velocities, and atomic numbers of modern science. Rather, it was a mystical belief, common in many ancient societies, in which various numbers represented things like love, gender, and hate. Even numbers were female while odd numbers were male. The number 1 was the omnipotent One and the generator of all numbers. The number 2 was the first female number and represented diversity. The number $3 = 1 + 2$ was the first male number composed of unity and diversity. The number $4 = 2 + 2$ was the number for justice since it is so well balanced. The number $5 = 2 + 3$ was the number of marriage since it is the union of male and female. The basic elements of antiquity, earth, air, water, and fire, were composed, Pythagoras believed, of hexahedrons, octahedrons, icosahedrons, and pyramids – geometric solids differing in the number of faces.

Before you make fun of this simplistic view of the world, consider how many people have "lucky" numbers or those who consider 13 to be unlucky or those who associate the number 18 with life, because the Hebrew letters for *life* have numerical equivalents which add up to 18. Numerology is alive and well even as we read! Pythagoras thought that the properties of whole numbers are worthy of study, and in that spirit, made important contributions to number theory.

He started by observing that some integers have many factors while others have relatively few. The factors of 12 are, for example, 1, 2, 3, 4, and 6. (He didn't consider the number a factor of itself, i.e., he considered only *proper* factors.) On the other hand, the proper factors of 10 are 1, 2, and 5. (Another word for factor is *divisor*, more common among mathematicians.) So Pythagoras decided to compare a number with the sum of its divisors. If the sum exceeds the number, he called it *abundant*, while if it is less than the number, he called it *deficient*. Since, $1 + 2 + 3 + 4 + 6 = 16$ which is greater than 12 (written $16 > 12$), 12 is abundant. As $1 + 2 + 5 = 8$ which is less than 10 (i.e., $8 < 10$), it follows that 10 is deficient.

What do we do with numbers like 6 and 28 which equal the sums of their proper divisors? ($6 = 1 + 2 + 3$ and $28 = 1 + 2 + 4 + 7 + 14$.) These were (and are!) called *perfect* numbers. In Euclid's thirteen-volume *Elements*, we find a method of finding perfect numbers (there don't seem to be a lot of them! – the next one after 28 is 496) which may have originated with the Pythagoreans. Before we get to that, let's look at an example.

Example 3-1 *Determine if the following numbers are deficient, abundant, or perfect. (a) 15   (b) 18   (c) 24*

(a) The proper divisors of 15 are 1, 3, and 5. Adding these gives us $1 + 3 + 5 = 9$. Since $9 < 15$, the number 15 is deficient.

(b) The proper divisors of 18 are 1, 2, 3, 6, and 9. Adding these gives us $1+2+3+6+9 = 21 > 18$ so 18 is abundant.

(c)The proper divisors of 24 are 1, 2, 3, 4, 6, 8, and 12. Adding these gives us $1+2+3+4+6+8+12 = 36 > 24$ so 24 is abundant. ∎

Now, let's get back to Euclid's method for finding perfect numbers. We begin by considering sums of "powers of two" beginning

with 1 – an interesting problem in its own right. The reader is re-
minded that the numbers 1, 2, 4, 8, and so on, obtained by starting
with 1 and then proceeding by repeatedly doubling the preceding
number are powers of two, that is, $1 = 2^0$, $2 = 2^1$, $4 = 2^2$, $8 = 2^3$,
and so forth. (That $2^0 = 1$ is a mystery! It follows from observing
that canceling 2's in the numerator and denominator of a fraction
leaves a 1.)

Let's look at the first few answers and look for a pattern. We
have

$$1 + 2 = 3$$
$$1 + 2 + 4 = 7$$
$$1 + 2 + 4 + 8 = 15$$
$$1 + 2 + 4 + 8 + 16 = 31$$
$$1 + 2 + 4 + 8 + 16 + 32 = 63$$
$$1 + 2 + 4 + 8 + 16 + 32 + 64 = 127$$

Notice that the sums 3, 7, 15, 31, 63, and 127 are each one less
than a power of two. In fact, the sum in any given equation is one
less than the next power of two. (In the last equation, 127 is one
less than the next power of two after 64, namely 128.) In the spirit
of the Greeks, let's prove this! Let's consider the sum, denoted S, of
the powers of two from 1 to a final power of two, which we will call
L (for last power of two). So,

$$S = 1 + 2 + 4 + \cdots + L \qquad (3.1)$$

Since L is the last power of two in the sum, $2L$ is the next power
of two after L. Thus, we want to show that $S = 2L - 1$. Now follow
this incredible argument. Even though we desire S, let us multiply
both sides of this last equation by 2. We get

$$2S = 2 + 4 + 8 + \cdots + L + 2L \qquad (3.2)$$

The reason the "L" is still present is because $L/2$ appeared just
before L in the original equation (3.1). Now get ready for the brilliant
step. Let us subtract the left and right sides of the first equation (3.1)
$S = \ldots$ from the corresponding sides of the second equation (3.2)
$2S = \ldots$. The left subtraction is $2S - S$, which is just S, right?

Twice something minus the something yields that same something. On the right we have common numbers which therefore cancel in the subtraction leaving just $2L - 1$. So we have $S = 2L - 1$, which is what we wanted to prove!

At this point, (if you didn't toss the book aside) you are probably wondering what all this has to do with perfect numbers. Looking at the sums 3, 7, 15, 31, 63, and 127, notice that the sums 3, 7, 31, and 127 are *prime* numbers – numbers with no proper divisors except 1, while 15 and 63 are not. It turns out that when the sum is a prime number, multiply the sum by the last number on the left side of the equation and you get a perfect number! So we get the first few perfect numbers like this:

$$3 \times 2 = 6$$
$$7 \times 4 = 28$$
$$31 \times 16 = 496$$
$$127 \times 64 = 8128$$

It is currently not known if there is an infinite supply of perfect numbers. Mathematicians have never found an *odd* perfect number, by the way. Is this because there aren't any or because they are so large that our computers can't get to them? (Yet!)

The Pythagorean brotherhood was entranced by mathematics. They believed that mathematics was the key to the nature of all things and that mathematics was everywhere.

Pythagoras noted that when we attempt to tile a floor with square tiles, we succeed because the meeting point of four right angle corners leaves no space, that is, four right angles add up to $360°$. Take a glance at Figure 3-7 to see through the eyes of Pythagoras.

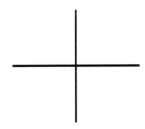

Figure 3-7

The next observation of Pythagoras was that six equilateral triangles meeting at a point also leave no space, as you can see from
Figure 3-8, which illustrates that six 60° angles also add up to 360°.
We are assuming that you recall that a triangle with three equal sides
is called *equilateral* and that its angles are also equal – implying that
each is 60°.

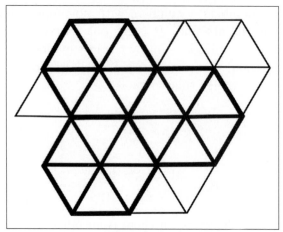

Figure 3-8

If you take an even closer look at Figure 3-8, you might notice
that we can view it as groups of six triangles which form hexagons –
six-sided polygons with equal sides and angles. Many old bathroom
floors are tiled this way. Bees are apparently well versed in geometry
since hexagons are found in honeycombs. Now equilateral triangles,
squares, and "regular" hexagons are examples of *regular polygons* -
polygons with equal sides and equal angles. Stop signs are regular
octagons, for example. Each interior angle of a stop sign is 135°. We
would, for this reason, not tile a floor with them. Put in mathematical terms, 135 is not a divisor of 360, while 60, 90, and 120 are. (The
interior angles of regular hexagons are 120°.) Thus, using triangles,
squares, or hexagons, we have found all the ways of placing tiles on
a floor, using regular polygons. (A polygon is, as its name suggests,
a figure with many sides.) This, of course, reinforced Pythagoras'
belief that "everything is number."[6]

Pythagoras noticed something else. If we pluck two strings of
equal thickness such that the ratio of their lengths is 2:1 (i.e., one of

---

[6]He might have said this if ancient Greek dentists administered Novocain.
Number – get it?

them is twice as long as the other), we will get pitches that are an octave apart. For those who are familiar with music theory, we could have "middle c" C1 and the next octave higher C2, for example. The shorter string has the higher pitch. Guitarists already know this. If we put a finger on the twelfth fret (*frets* are the little metal rods placed along the neck of the guitar), which is located exactly half way down the neck, thereby shortening the string by half, we get a note one octave above the "open" string's note. This principle applies to all stringed instruments.[7]

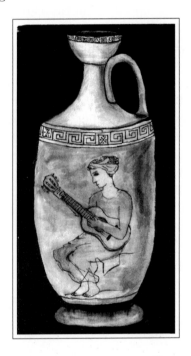

Pythagoras didn't stop with the octave. He considered two strings of lengths with ratio 3:2, (i.e., their lengths might be 15 inches and 10 inches, since $15/10 = 3/2$) and found that the result was the pleasing or consonant interval called a *fifth*. Musicians measure the distance between two notes by starting the scale on the first of the notes and count up to the second note. The notes C and G form a fifth in light of the sequence of consecutive notes in the C scale – C, D, E, F, and G. The ratio 4:3 yields the interval called a *fourth*, that is, a C to an F. He found that the pleasing musical intervals already known in music corresponded to ratios with small numbers. They constitute

[7]But we won't harp on this anymore.

the basis of harmony in the musical tradition of Western civilization. Once again, we can hear Pythagoras smugly assert that everything is number, perhaps in a singsong voice.

Some say that Pythagoras did all of this not with strings but with hammer blows on metal. The principle is the same whether we use the length of a column of vibrating air in a flute, the length of a vibrating string, or the tension in a drum skin. We used strings here because of the popularity of the guitar.

Oddly enough, since the planets move across the sky at different speeds, that is, their speeds form ratios relative to one another, the Ancient Greeks believed that there were musical intervals associated with them. They referred to this as the *music of the spheres.* Medieval scholars wrote of these celestial harmonies.

It is interesting to note that the Greek educational curriculum consisted of four subjects: geometry, astronomy, music, and arithmetic (by which they meant number theory) and considered mathematics the common base of these topics – all interrelated. These subjects were renamed the *quadrivium* by Roman scholars and after being augmented by logic, oratory, and rhetoric, were once again renamed the seven liberal arts – the curriculum of the Middle Ages.

Before we leave Pythagoras, we present an algorithm (procedure) similar to the method he used to generate Pythagorean triples. He may have obtained this from the Babylonians. Take two numbers p and q that satisfy three properties:

1. $p > q$,

2. p and q have different *parity* (i.e., one is even and the other is odd), and

3. p and q have no common divisor except 1.

Now we find a, b, and c as follows:

$$a = p^2 - q^2$$
$$b = 2pq$$
$$c = p^2 + q^2$$

The numbers p and q are called *generators.* As an example, let's consider:

Example 3-2 *Find the Pythagorean triple for the generators $p = 2$ and $q = 1$.*

Using the equations for a, b, and c, we get $a = 2^2 - 1^2 = 3$, $b = 2 \times 2 \times 1 = 4$, and $c = 2^2 + 1^2 = 5$. Wow! We get the beautiful triple $3, 4, 5$. ■

Let's do that again.

Example 3-3 *Find the Pythagorean triple for the generators $p = 3$ and $q = 2$. (We can't let $q = 1$, because of condition 2 above.)*

Using the equations for a, b and c we get $a = 3^2 - 2^2 = 5$, $b = 2 \times 3 \times 2 = 12$, and $c = 3^2 + 2^2 = 13$. Amazing! This is the famous $5, 12, 13$ triple. ■

At this point, let us assure you that a tradition of contemplating mathematical truths for their own sake was firmly established. Pythagoras and his followers did not develop mathematics to design a can opener or measure a field for the county tax assessor. The Ancient Greek mathematicians were *philosophers* – lovers of knowledge, as the term denotes. They looked down on the cabinetmaker or the smith – on those who dealt with the grubby day-to-day chores of laborers. Many intellectuals share this attitude today. Consider the age-old "theory versus practice" debate.

In Plato's day, the focus of philosophy was on all-abiding, eternal truths – not truths about this particular table or that particular farmhouse. In ten years, the table we are writing on and the room we are sitting in might not even exist. Mathematical truths like $2 + 2 = 4$, on the other hand, will live forever! More on this later.

Early Greek mathematicians shared a naive belief that starting with integer lengths like 7 and 38 and then subdividing them into fractions like $\frac{7}{3}$ and $\frac{38}{9}$, they could express any length. We call such quantities *rational* numbers, because they are ratios of integers. Some examples include $3\frac{1}{2} = \frac{7}{2}$, $0.13 = \frac{13}{100}$, and $2.4 = \frac{24}{10} = \frac{12}{5}$.

Imagine their surprise when they realized that this cannot be done for some numbers, that is, some numbers are *irrational*. The first example that the Pythagoreans encountered was the hypotenuse of a right triangle whose legs are both 1. By the theorem, the hypotenuse is $\sqrt{2}$. The Greeks would have called these lengths 1 and $\sqrt{2}$ *incommensurable*, meaning that they cannot equal the same length multiplied by (different) whole numbers. While $\sqrt{2}$ can be estimated to any desired degree of accuracy using rational numbers, such as 1.414 (that is, $1414/1000$), it is irrational! Before we prove this, we must cover a few basics.

1. The ratio of two integers can always be reduced to lowest terms. $\frac{6}{8}$, for example, can be reduced to $\frac{3}{4}$, which is said to be in lowest terms, that is, the numerator and denominator have no common divisor to cancel.

2. The square of an even number is even, while the square of an odd number is odd. Another way to say this is that squaring preserves the parity of a number. Consider the squares of the first few odd numbers 1, 3, 5, and 7. They are 1, 9, 25, and 49. They are all odd.

3. The ratio of two odd numbers may or may not be in lowest terms. Consider $\frac{15}{31}$ which is and $\frac{27}{33}$ which is not in lowest terms. $\frac{27}{33} = \frac{9}{11}$. On the other hand, the ratio of two even numbers is never in lowest terms. It may always be reduced because we can cancel a 2. Consider $\frac{18}{20}$. This becomes $\frac{9}{10}$ upon cancellation of the 2's.

Here we go with the proof that $\sqrt{2}$ is irrational. We will use a very popular method of proving things known as a *"proof by contradiction,"* where we assume that the statement we are trying to prove is false and then arrive at a contradictory conclusion. In drawing such a conclusion, we deduce that the original assumption must be false. Therefore, we will assume that $\sqrt{2}$ is a rational number and it is $\frac{a}{b}$, reduced to lowest terms, that is, a and b have no common divisor. (There is no harm in this because if it is not in lowest terms, just reduce it until it is.) We will show that this is impossible by obtaining a contradiction. If $\sqrt{2} = \frac{a}{b}$, then $2 = \frac{a^2}{b^2}$ (obtained by squaring both sides). Rewrite this as $2b^2 = a^2$, which implies that a^2 is even – it is 2 times something. Then by (2) above, a must also be even, implying that a is divisible by 2. Then it is 2 times something, that is, $a = 2m$ for some integer m. Then $a^2 = (2m)^2 = 2m2m = 4m^2$.

Let's substitute $4m^2$ for a^2 in the equation $2b^2 = a^2$ several lines above, yielding $2b^2 = 4m^2$. Dividing both sides by 2 yields $b^2 = 2m^2$. Then b^2 is even and so is b.

Where are we then? It seems that both a and b are even. But didn't we say that the fraction $\frac{a}{b}$ was reduced to its lowest terms? This is impossible in light of (3) and we have obtained our contradiction. Yes! Thus, the original assumption – that $\sqrt{2}$ was rational – must be false. We have an irrational number, $\sqrt{2}$.

The Pythagoreans tried to keep this a secret, and it is alleged that they drowned one of their members who revealed it to the outside world. In any event, the story leaked out and the Greeks were in shock. It is largely because of this that they focused on geometry, almost to the exclusion of computing. The irrationality of a quantity or of a length was no longer very important and could be swept under the rug. (They were wrong in this regard and merely postponed their troubles to a later time.) More advanced texts on the history of mathematics tell of the attempts of Eudoxus to make sense out of line segments of irrational length. Many problems were not successfully dealt with till the nineteenth century (A.D.!).

We close this discussion of Pythagoras with a proof of his famous theorem. To date, there are over 250 different proofs of the theorem. In fact, in 1876, Congressman James Abram Garfield, who later became the twentieth president, published his own proof of the Pythagorean Theorem. Fear not. We shall present only two.

Consider the square with vertices A, B, C, and D in Figure 3-9. We add points E, F, G, and H in just the right places so that the line segments they determine have the labeled lengths a and b. We then draw line segments $\overline{EF}$, $\overline{FG}$, $\overline{GH}$, and $\overline{HE}$ which are equal in light of the fact that the four right triangles in the figure are congruent and have, therefore, the corresponding sides of equal length, called c.

Reminder: One of the ways to show that two triangles are *congruent*, (i.e., one can be superimposed on the other with their respective vertices coinciding) is to demonstrate that two sides and the included angle in both triangles are equal. This criterion is abbreviated *SAS* for *side, angle, side*. The A is placed between the two S's to indicate that we require the included angle. And, of course, to avoid writing an impolite word.

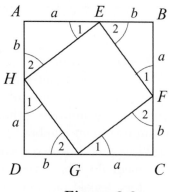

Figure 3-9

Now consider the acute angles of the four right triangles labeled $\angle 1$ and $\angle 2$. We need only two angles by reason of the congruence of the triangles. Observe, furthermore, that $\angle 1 + \angle 2 = 90\,°$, since the angle sum of a triangle is $180\,°$ and the third angle is $90\,°$. Since the sum of the three angles at each of the vertices E, F, G, and H form a straight angle ($180\,°$), it follows that the unlabeled angles are $90\,°$, implying that quadrilateral $EFGH$ is indeed a square. Observe that it did not suffice merely to assert the equality of its four sides, because then that would only have shown it was a rhombus.

To finish the proof, we shall compute the area of the larger square in two different ways and then equate the answers. The first way is obvious. Since a side of the square is $a + b$, the area is $(a + b)^2$ which is $a^2 + 2ab + b^2$. (If your algebra is a bit rusty, just multiply $a + b$ by itself.)

The second way is to add up the pieces. The inner square $EFGH$ has area c^2. The four right triangles each have area $\frac{1}{2}ab$, and since there are four of them, they contribute $2ab$, finally yielding $c^2 + 2ab$. Let's equate the two answers:

$$a^2 + 2ab + b^2 = c^2 + 2ab$$

which, after subtracting $2ab$ from both sides, gives us the Pythagorean Theorem.

We shall include another proof of this theorem to demonstrate the exciting idea that mathematics is a highly creative endeavor. Many of us, in elementary school, had teachers who insisted that we did the problems their way – always! Mathematics is often taught in a study-memorize-regurgitate manner which doesn't permit students to see how creative they can be.

Before we begin this proof, let's recall what it means for two triangles to be similar. They can have vastly differing sizes, but they must have the same shape. Well, what determines the shape of a polygon? Two things:

(a) They must have the same angles.

(b) Their sides must have the same ratios.

These conditions may exist separately, in which case we do not have similarity. Consider two rectangles such that the first is 2 by 5 and the second is 4 by 7. Condition (a) is met, that is, all rectangles have the same corresponding angles, while the ratios are unequal, that is, $5/2$ is not equal to $7/4$. Alternatively, one can say that we enlarged the width of the smaller rectangle from 2 to 4 (we doubled it) while we enlarged its length from 5 to 7 (a mere 40% increase!). If you enlarged the first rectangle to increase the scale of your drawing and you wanted the width to double, then you would wind up with a 4-by-10 rectangle (which is similar to the 2-by-5 one).

On the other hand, consider changing the 2-by-5 rectangle to a 2-by-5 parallelogram with base angles, say, $80°$ and $100°$, which is the same as tilting the sides by $10°$. We then obtain a quadrilateral with a different shape.

Now the amazing thing about triangles is that either one of the two conditions above suffices to establish similarity, that is, condition

(a) implies (b) and vice versa! If the angles of one triangle equal the corresponding angles of the other, bingo – they are similar, and we can rest assured that their sides are proportional. But it gets better! Since the sum of the angles of a triangle is 180°, if two angles of one triangle equal two angles of another, the third angles must also be equal and the triangles must be similar. Give us two right triangles which both contain a 30° angle and we shall know at once that they are similar, without having to add 90° and 30° and then subtract the sum, 120°, from 180° and get 60°, the third angle of both triangles. Actually, the two acute angles of any right triangle add up to 90° and are called *complementary*. So the 60° would be calculated at once by realizing that it's the complement of 30°.

Now on with the proof. In right triangle $\triangle ABC$ of Figure 3-10, we see the so-called altitude CD which is *perpendicular* to (makes a 90° angle with) hypotenuse AB.

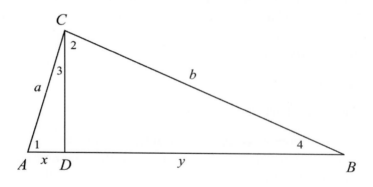

Figure 3-10

The segments AD and DB have lengths x and y which add up to c, the length of the hypotenuse. Now the angles labeled $\angle 1$ and $\angle 2$ are equal because they have the same complement, namely the angle labeled $\angle 3$. In the same way, the angles labeled $\angle 3$ and $\angle 4$ are equal. Then all three right triangles in the figure are similar! They all have the same acute angles. Then we have a right to assume that certain ratios are equal. In particular, by comparing triangles $\triangle ACD$ and $\triangle ACB$, we get $a/x = c/a$. Then we compare triangles $\triangle BCD$ and $\triangle BCA$, getting $b/y = c/b$. Each of these equations may be simplified by "cross multiplying." The first becomes $a^2 = cx$ and the second becomes $b^2 = cy$. If we add the left and right sides of these equations, we get $a^2 + b^2 = cx + cy$ which factors into $c(x + y)$. Since $x + y = c$, the last expression can be written as $c(c)$ or just c^2. We are finished! This proof is a masterpiece of reasoning and illustrates the inevitable succession of logic in a (valid) mathematical proof. The Greeks admired this sort of logic and thought of the study and creation of geometry as ends in themselves. It was also, needless to say, an excellent educational tool.

The next great hero of Ancient Greek geometry is Hippocrates of Chios, not to be confused with the namesake of the Hippocratic Oath. Hippocrates was bent on deriving the area of a *lune* – a region bounded by arcs of two circles. The moon, at times, has this shape, which accounts for this strange name. The lune of Figure 3-11 is obtained by drawing a semicircle with center at O and radius AO, of length 1. It follows that AB, the diameter, has length 2. So far so good. Now we draw radius OC such that it is perpendicular to AB. Let's throw in AC for good measure. Then triangle AOC is a right triangle with legs of length 1, in which case hypotenuse AC has length $\sqrt{2}$. (This triangle was a thorn in the backsides of the Pythagoreans, but it returns as a friend here.)

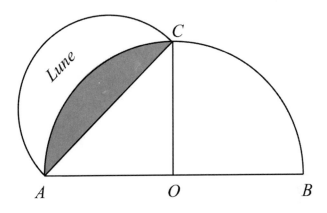

Figure 3-11

Notice that we have also drawn a semicircle whose diameter is AC. This semicircle is smaller than the original one, whose diameter has length 2. As noted above, AC has length $\sqrt{2}$ which is approximately 1.414. Now the good geometers of that time made a remarkable observation concerning the areas of similar figures which is highly relevant here since all semicircles are similar – they have the same shape.

Two similar polygons have a *linear magnification ratio* – the number that we must use to multiply the sides of one polygon to get the sides of the other. If we are given two similar rectangles, the first with sides 2 and 5, and the second with sides 6 and 15, the linear magnification ratio is 3, that is, each side of the larger rectangle is three times as long as the corresponding side of the smaller. On the other hand, the areas of the two rectangles are 10 and 90. The ratio of area magnification is 9, not 3. What's going on here? Why do we get the square of the linear magnification ratio?

Let's do this abstractly. Suppose the smaller rectangle has dimensions a and b, and the larger one has dimensions ka and kb, where we are assuming that k is the linear magnification ratio. Then the area of the larger rectangle is k^2ab which shows us that the area magnification ratio is the square of the linear magnification ratio. The Greeks extended this observation to other figures as well, including the circle.

In Figure 3-12, we consider the two semicircles with diameters AB and AC of the previous figure. Since their diameters have lengths 2 and $\sqrt{2}$, the area magnification ratio is $\left(\frac{2}{\sqrt{2}}\right)^2$, which after squaring the numerator and denominator and simplifying yields 2.

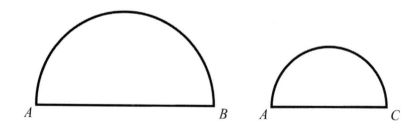

Figure 3-12

Then the larger semicircle has twice as much area as the smaller one, implying that half the larger has the same area as the smaller, as indicated by Figure 3-13, in which we equate the areas of the semicircle with diameter AC and the quarter-circle (AOC).

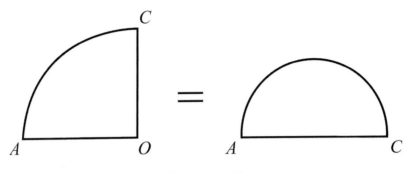

Figure 3-13

Now in the original diagram, the semicircle and quarter-circle overlap in the shaded segment with corners at A and C. If we remove this overlap from the semicircle and quarter-circle, the leftovers must have the same area. But the leftovers are the lune with corners at A and C and the right triangle AOC. Since the base and height of this right triangle have length one, its area is $\frac{1}{2}$. Then this is also the exact area of the lune! (Wow)^2! Please bear in mind that this magnificent argument is made even more magnificent when one realizes that Hippocrates did this in approximately 450 B.C. This is the first exact calculation of the area of a *curvilinear figure* (one with curved sides). Understand that this is about 200 years prior to the discovery, by Archimedes, that the area of a circle is πr^2.

The fifth-century philosophers Zeno and Democritus[8] wrestled with a problem that would not be completely solved for another two thousand years: the problem of infinitesimal magnitudes. Zeno argued that motion was impossible! An object attempting to get from point A to point B would first need to get to the midpoint of its journey; let's call this point C. But before the object reached point C, it would have to get to the midpoint between A and C. Once again, let's call this new point D. This argument may be repeated ad infinitum, from which Zeno concluded that motion was impossible. It would require traversing infinitely many points in a finite amount of time.

Democritus said motion was possible by positing the existence of ultimate indivisible particles, called atoms, out of which all things

---

[8]As Democritus would say, "A life without festivity is a long road without an inn."

are constructed. He asserted that one couldn't continue to subdivide something indefinitely.

Mathematicians today, like us, understand that a finite quantity can be represented as a sum of infinitely many progressively smaller quantities. An easy example of this is given by the infinite, repeating decimal $0.999\ldots$ which clearly must equal 1. If you believe it is less than 1, you must tell how much the deficit is, and when you do, we can show you a large enough chunk of (finitely) many 9's in this decimal which are closer to 1 than the deficit. Gotcha!

1. In the accompanying figure, straight lines $\overline{AB}$ and $\overline{CD}$ intersect at E. The angles at C and D are right angles. Prove that $\angle A$ and $\angle B$ are equal using the fact that lines $\overline{AC}$ and $\overline{DB}$ are parallel. (How do we know they are parallel?) Now prove that $\angle A$ and $\angle B$ are equal without using the parallelism of lines $\overline{AC}$ and $\overline{DB}$.

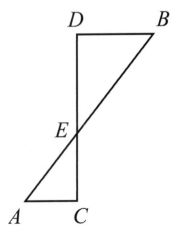

2. If the angles of a triangle are in the ratio of 1:2:3, what is the measure of the smallest angle?

3. Prove that the sum of the angles of a quadrilateral (four-sided figure) is $360°$. Why is this fact obvious for rectangles?

4. Two complementary angles are in the ratio of 1:4. Find the measure of the smaller angle. Remember, complementary angles add to $90°$.

5. What is the angle sum of a pentagon? (*Hint*: Add two diagonals to create three triangles.)

6. Generalize Exercise 5 to polygons with n sides and show that the sum of the angles is $(n-2) \times 180°$. Does this agree with the fact that the angle sums of triangles and quadrilaterals are $180°$ and $360°$, respectively?

7. *Amicable numbers* are a pair of numbers in which the proper divisors of one add to the other and vice versa. In numerology, amicable numbers represented friendship. Show that 220 and 284 are amicable numbers.

8. Prove that the square of an even number is even while the square of an odd number is odd. (*Hint*: $n = 2k$ is an even number for all k and $n = 2k + 1$ is odd.)

9. Use the procedure outlined in this chapter to find the Pythagorean triples from the given generators.
   (a) $p = 4$ and $q = 3$ (b) $p = 5$ and $q = 2$
   (c) $p = 4$ and $q = 1$ (d) $p = 7$ and $q = 4$
   (e) $p = 7$ and $q = 2$ (f) $p = 5$ and $q = 4$

10. Use the procedure outlined in this chapter to find five sets of Pythagorean triples whose numbers are larger than 100.

11. Using the fact that the square of a multiple of three (i.e., a number with three as a divisor, like 6 or 15) is again a multiple of three, while the square of a nonmultiple of three (like 8 or 10) is not a multiple of three, prove that $\sqrt{3}$ is irrational. Model your proof after the proof of the irrationality of $\sqrt{2}$.

12. **Numerology**. The name of a person, place, or thing can be changed into a number and that number can reveal the character of the thing. Using the tables below we can change, for example, the date 7/4/1999 into $7 + 4 + 1 + 9 + 9 + 9 = 39$

which can be simplified to $3 + 9 = 12$ and again to $1 + 2 = 3$, showing us that it is a day to enjoy life, or the name Thales into $2 + 8 + 1 + 3 + 5 + 1 = 20$ and again into $2 + 0 = 2$, showing us that Thales was a well-balanced and well mannered person.

1	2	3	4	5	6	7	8	9
A	B	C	D	E	F	G	H	I
J	K	L	M	N	O	P	Q	R
S	T	U	V	W	X	Y	Z	

Number	Meaning
1	Ambitious, strong willed, courageous
2	Balanced and well mannered
3	Versatile, gregarious, confident
4	Honest, reliable, stable
5	Adventurous, enthusiastic, independent
6	Sincere, successful, leadership qualities
7	Imaginative, poetic, naturally talented
8	Demanding, strong, financial success
9	High achiever, creative ability

Do this for the names

(a) Euclid   (b) Hypatia
(c) Pythagoras (d) Euler

Suggestions for Further Reading

1. Boyer, Carl. *A History of Mathematics*, John Wiley & Sons, New York, 1991.

2. Clawson, Calvin C. *Mathematical Mysteries: The Beauty and Magic of Numbers*. Perseus Press, New York, 2000.

3. Heath, Thomas L. *History of Greek Mathematics: From Thales to Euclid*. Dover, New York, 1981.

4. Scott, J. F. *A History of Mathematics; from Antiquity to the Beginning of the Nineteenth Century*. Taylor & Francis, London and New York, 1969.

5. Smith, D. E. *History of Mathematics*. Dover, New York, 1958.

6. Solow, Daniel. *How to Read and Do Proofs*. John Wiley & Sons, New York, 1990.

Chapter 4

Greeks Bearing Gifts

A great concern among Greek philosophers, most notably Plato and Aristotle, was the meaning of universals, or forms. When a person uses the word dog, to what do they refer? In the famous

quote, "Outside of a dog, a book is man's best friend. Inside of
a dog, it's too dark to read..." the form or universal dog is being
discussed. Some dogs are tall, others short. They occur in a variety of
colors. Some are friendlier than others. Yet they all share something
or they wouldn't be dogs. The Greek philosophers were interested
in the abiding essence of dog.[1] After all, every dog on this planet
will eventually die – but "dogness" will go on forever, we hope. This
is called the problem of universals in philosophy and it has many
different answers.

   The famous fourth-century (B.C.) philosopher Plato believed
that there is a perfect dog which resides in the world of universals
– a mystical place where our immortal souls come from prior to our
birth. This universal dog is stripped of all of its nondog-specific characteristics. It has no specific color or height or weight. It wouldn't
be your pet, Fango, because he is a specific dog, while the universal
dog has only general characteristics essential to every dog. Plato's
perfect puppy will never die – and it doesn't have to be walked. It
is truly the ideal dog. In fact, the term "ideal" comes from Plato's
world. The forms there give us the "idea" of dog, cat, man, or what-
ever class of things we have the idea of. The forms are eternal and

---

[1]A name for a cologne?

worthy of our study, whereas earthly objects are imperfect shadows of the forms. You may have read Plato's cave analogy in his book *The Republic*, in which he expounds on this theme. It should come as no surprise that Plato loved geometry. Geometry establishes necessary connections between the forms of the polygons we see around us. We don't speak of *this* rectangular table or *that* circular clock on the wall in geometry. We discover truths, rather, of the perfect circle or the perfect rectangle. These are eternal truths.

Plato founded a school of philosophy which he named the Academy, and legend has it that he wrote over its portals "Let none enter here who have not studied geometry." Geometric truths, thought Plato, are apprehended through pure reason, unsullied by perception. The diagrams simply allow us to recollect things our souls knew in that other world (of universals). A legacy of his is "Platonic Love" – pure and chaste – a love of the eternal soul of the other person unblemished by carnal (and temporary) lust for the flesh. You can imagine that Plato wasn't the life of the party.

Aristotle, a student of Plato's, departed from all of this. He rejected the world of universals, maintaining that the essence of dog resides in each dog. Perception, without reason, however, will not reveal the essence. Rather, we form the idea of dog by inducing it from observation of many particular dogs.

Their differences aside, these great philosophers, including Plato's teacher Socrates, secured the firm belief in logical structure employing classes with certain properties. Once we establish that something is a member of a class, we may assume that it has all of the properties of that class. So if Fango is a dog, then he barks, wags his tail, and chases chariots.

Aristotle posited that knowledge begins with certain basic irrefutable assumptions or axioms, without which nothing can be said. If one were to say to him, "You must prove everything – even your most basic assumptions," he would have disagreed by showing you an absurd "*infinite regress*" that would result. Statement A would depend on B, which would depend on C, and so on, to infinity, and nothing would be proven!

Aristotle, by the way, was fascinated by infinity and wrote that a line segment is infinitely divisible, disagreeing with the views of pre-Socratics like Zeno and Democritus who believed that everything has an ultimate building block or atom which is indivisible. Zeno stated several paradoxes of motion resulting from his belief that a moving

1/3 of the Nine Muses

object travels from one point of its trajectory to the next. We know today that between any two points on a line, there are infinitely many others.

The logical foundations of mathematics would have to wait for the great Euclid of Alexandria. Why Alexandria if he was Greek? Alexander the Great (the great what?) conquered an enormous territory including Greece, the Balkans, Turkey, Egypt, the Middle East, Persia, and so forth, extending all the way to northern India. He founded a city named Alexandria where a great Museum flourished. It contained a library of over 100,000 Greek manuscripts. Upon Alexander's death in 323 B.C., his vast kingdom was split into three empires, one being Egypt. Though the Romans conquered Egypt in the first century B.C., they permitted Greek culture, including the Museum, to flourish. The library was finally burned to the ground in 641 A.D. by Islamic invaders. The word "museum" derives from the Muses, the nine sisters of the arts, from which "music" is also obtained.

Euclid[2] flourished around 300 B.C. and lectured at the Museum. He wrote a thirteen-volume work called *The Elements*, which has appeared in more editions than any other book except for the Bible.

The Elements summarizes 300 years of Greek geometry and num-

---

[2]When Euclid was asked by a student, "What shall I gain by learning these thing?" he replied, "Someone give him a penny, since he must make a profit from the things he learns."

ber theory – but it does much more than that! Euclid establishes definitions. A subject must start by defining the things it studies. A point has location without extension. A line is the shortest distance between two points. Parallel lines don't meet no matter how far they are extended. A triangle is a three-sided figure, and so forth and so forth.

He then presents common notions and axioms, such as "equals added to equals yield equals" and "the whole is greater than its parts." He assumes, for example, that through a point not on a given line, there exists a unique (only one) line parallel to the given line, though he states this in a very roundabout way. This "parallel postulate" has been doted on for centuries until, in the nineteenth century, new geometries were founded without this postulate. They are called *non-Euclidean* and play an important role in the theory of relativity.

Euclid's volumes were so well done and summarized earlier works so thoroughly that it made older works obsolete. Later-day Islamic translators felt it unnecessary to translate these works, resulting in the tragic loss of many earlier manuscripts.

Euclid was interested in prime numbers and proved that there are infinitely many of them. In case you think this is obvious, consider this. After the prime number 2, no even number is prime. This eliminates half the numbers. After 3, every third number is a multiple of three and, therefore, not a prime. After 5, every fifth number fails to be prime, and one sees that it seems quite conceivable that at some point there are no more primes. Before we look at Euclid's proof of the infinitude of primes, we need some preliminary observations:

1. Every integer can be factored down to prime factors. Let's look at an example of this fact.

Example 4-1 *Write 100 as the product of prime factors.*

Since $100 = 4 \times 25$, which are not prime factors, keep factoring. $100 = 2 \times 2 \times 5 \times 5$, and we get our prime factorization! This is normally written using a *factor tree*

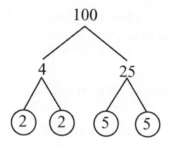

and exponential notation $100 = 2^2 5^2$. ∎

2. A number is divisible by another number if and only if the second number is a factor of the first. Thus, 36 is divisible by 12 because $36 = 3 \times 12$.

3. If a number, say n, is divisible by another, say m $(m > 1)$, then if we add one to the first number, it is no longer divisible by the second, that is, $n + 1$ is **not** divisible by m.

4. A (rather lengthy) product of many integers is divisible by each of them. For example, $2 \times 3 \times 5 \times 7 = 210$ which is divisible by each of 2, 3, 5, and 7. (It is divisible by more numbers such as 6 and 10.)

5. Finally, if a number > 1 is composite (not prime), it must have a prime factor.

Euclid began his proof by assuming that there are only finitely many primes and hoped to obtain a contradiction. If there are finitely many primes, let us call them a, b, c, and so on, up to the last (or greatest) prime, say, p. Then any number larger than p is, by our assumption, a composite number – because we are assuming that p is the biggest prime. Now, said Euclid, consider the number $a \times b \times c \times \cdots \times p$ called N, and finally consider the even larger number $N + 1$. Since this number is obviously bigger than p, it is a composite number. Then it must have a prime factor. This prime is clearly a factor of N and therefore goes into N. It follows, then, that this prime factor does not go into $N + 1$. This is an enormous contradiction. $N + 1$ is a composite number without a prime factor! If you don't get it, read the proof again – it's worth it. This is one of the greatest proofs of antiquity.[3] It is even more amazing to us

---

[3]Scholars are not sure where "antiquity" was located.

that Greek scholars asked the question, "are there infinitely many primes?" – this question has very little bearing on the price of tea in Babylon. Once again, we see that the Greeks loved knowledge for its own sake and not for its technological benefits.

Euclid developed an algorithm (procedure) for determining the *greatest common divisor*, henceforth abbreviated GCD, of two integers. Reminder: The GCD of 15 and 20 is 5 because 5 is the greatest number which goes into both 15 and 20. It is easy to figure out the GCD of small numbers by guesswork, but what do we do if the numbers are large? Euclid began with the following assumption. If a number n goes into x and into y, then it also goes into $x - y$. He then reasoned that we could subtract the smaller number from the larger as many times as possible and then compute the GCD of the remainder and the smaller number. This yields a much easier problem, after which the procedure may be repeated. For example,

Example 4-2 *Find the GCD of 30 and 650.*

Observe that we may subtract 30 (many times) from 650 until we get a remainder of 20. (A quick way to do this is to divide 30 into 650, discard the quotient, 21, and keep the remainder of 20.) So, $650 \div 30 = 21$ R 20. Then the GCD of the original pair, 30 and 650, must go into 20. Since it also goes into 30, we can simply compute the GCD of 20 and 30, which by guesswork is 10. Congratulations! This is the GCD of the original pair. If guesswork were inadequate here, we would repeat the procedure by dividing 20 into 30 and retaining the remainder of 10. The GCD of 20 and 10 is 10, which we already know is the answer. ∎

Example 4-3 *Find the GCD of 48 and 360.*

Once again divide the larger number by the smaller number and discard the quotient, $360 \div 48 = 7$ R 24, so we discard the 7 and keep the 24. We do this because the GCD of 48 and 360 is the same as the GCD of 24 and 48. So, divide again, if you don't see the answer yet, and we obtain $48 \div 24 = 2$ R 0. When we get a remainder of zero, the algorithm stops. When this happens, the answer is the last divisor, in this case 24. ∎

This is a clever algorithm and is as valid today as it was 2300 years ago. It is known as the *Euclidean algorithm* or the divisor algorithm.

Let us now turn our attention to a typical theorem in *The Elements*. Let us prove that adjacent angles of a parallelogram are *supplementary*, that is, they add up to 180°. In the drawing of Figure 4-1, we have extended side AB to an arbitrary point E.

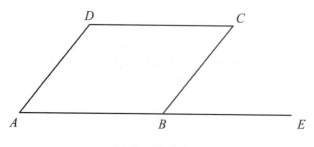

Figure 4-1

Then since AE is a transversal across parallel lines AD and BC, we have the equality of $\angle DAE$ and $\angle CBE$. Moreover, the two angles at vertex B, namely $\angle ABC$ and $\angle CBE$, satisfy $\angle ABC + \angle CBE = 180°$, because ABE is a straight line. Then, by the equality of $\angle DAE$ and $\angle CBE$, we can replace the latter by the former in the previous equation, yielding $\angle ABC + \angle DAE = 180°$. This is exactly what we had to prove. Note how every statement in this proof can be defended with previously proven theorems or by an axiom (or some irrefutable initial assumption like the fact that two angles which share a vertex and a side and whose nonshared sides form a straight line, are supplementary). Very few subjects, if any, share this meticulous logical structure – a pyramid rising on a clearly stated foundation. To be sure, errors have later been uncovered, but the idea remains today. Each branch of mathematics has a succession of axioms, definitions, and theorems.

It is a tribute to Euclid that we speak even today of Euclidean and non-Euclidean geometry.

Throughout the Hellenistic period, we find that the great Greek mathematicians flourished outside of Greece! Our next mathematical hero, Archimedes (287-212 B.C.), lived in Syracuse, a Greek settlement on Sicily (as the island is known today). He is remembered as an eccentric thinker who was, contrary to the spirit of Greek mathematics, interested in machines and other worldly objects. He is said to have invented a host of weapons used to defend Syracuse against the occasional Roman attack during the Punic Wars (between Rome and its mighty North African competitor, Carthage).

Having discovered the principles of the fulcrum, or in common parlance, the seesaw, and the compound pulley, he created huge pulleys with enormous metal claws which, when lowered from the fortified cliffs of Syracuse, hoisted Roman ships out of the water to great heights only to smash them to smithereens on the rocks below. These pulleys were easy to manipulate because they were compound pulleys – they had several moving wheels. In fact, these pulleys replaced hundreds of slaves in pulling friendly ships to dock in the ports of Syracuse. Archimedes expressed the laws governing these in mathematical terms.

The seesaw in Figure 4-2, illustrates his law that the first distance times the first weight equals the second distance times the second weight, or $W \times D = w \times d$. We are using caps for the left object's weight and distance and lowercase for the right. If the second weight is very small but its distance is sufficiently large, and if the first distance is very small, the first weight can be quite large. Hence a child on a seesaw sitting far away from the middle can counterbalance a heavy adult sitting closer to the middle. This principle gives us the leverage we need when we lift a heavy car using a jack. Can you picture Archimedes changing a chariot wheel this way?

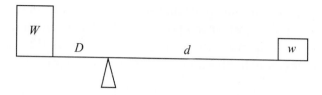

Figure 4-2

Archimedes[4] supposedly met his end when the Roman armies of Marcellus overran Syracuse. A soldier was sent to fetch Archimedes, who was in the middle of a lengthy geometric calculation in the sand. Without looking up, he told the soldier to wait until he finished. The enraged soldier then slew him. There is in this story an interesting contrast of cultures – Rome did not produce one truly great mathematician! But Romans could fight.

One of the greatest achievements of the immortal Archimedes was his conquest of the circle. His approach was to view it first as a square – a preposterous idea if there ever was one. It makes one recall the square wheels of cartoon caricatures of early transportation. If the "inscribed" square of Figure 4-3 has sides of length b, and if the distance from the center of the circle to the midpoint of each side has length h, then by thinking of the square as composed of four triangles, it has an area of $4 \times \left(\frac{1}{2} \times b \times h\right)$, that is, the area is four times the area of each triangle, using the Egyptian formula for the area of a triangle. Now the product of these numbers can be rearranged and recombined to yield $\frac{1}{2} \times (4 \times b) \times h$.

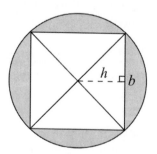

Figure 4-3

Note that $4 \times b$ is the perimeter of the square. It is a simpler expression than $b + b + b + b$! Let's call the perimeter P and the area

[4]Archimedes said, "Give me a place to stand and I can move the world!"

A and rephrase the previous equation for the area of the inscribed square: $A = \frac{1}{2}Ph$. (Recall that xy is shorthand for $x \times y$.)

The next step Archimedes took was to double the number of sides to yield an inscribed octagon. This requires four more points of contact with the circle, as shown in Figure 4-4, in which the lengths of the sides and distances from the center to the midpoints of the sides have changed and are now denoted by b' and h', respectively.

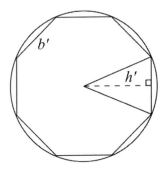

Figure 4-4

Observe that the area of the inscribed octagon (denoted A'), obtained by viewing it as eight triangles, is $8 \times \left(\frac{1}{2} \times b' \times h' \right)$. This can be rearranged to yield $A' = \frac{1}{2} \times (8 \times b') \times h'$, or just $A' = \frac{1}{2}P'h'$. Of course, P' is the new perimeter. Note, however, that the formula is the same! Realizing that P' and h' are bigger than P and h, Archimedes wondered what would happen if this doubling process could be continued forever – or *ad infinitum* (Latin for "to infinity"). What an exciting and tremendously sophisticated idea. Mathematicians are still grappling with the Greek idea of infinity today.

Archimedes saw that the perimeters of these figures get closer to the "limiting" value C which is the circumference of the circle, while the distances from the center to the midpoints of the sides approach r, the length of the radius of the circle. He discovered the beautiful formula $A = \frac{1}{2}Cr$, in which A now represents the exact area of the circle. Of course, he didn't stop here because he wasn't sure how to calculate C. You can't just put a ruler around a circle's circumference.

Now all circles have the same shape, that is, they are similar. Then the ratio of the circumference of any circle to its diameter is constant. We learn this in elementary school! The ratio is now called

π. Calling the diameter D, we have $C/D = \pi$, or $C = \pi D$. This can be changed to $C = \pi \times 2r$, since the diameter is twice as long as the radius. This is usually written $C = 2\pi r$ in our textbooks.

If we substitute the expression $2\pi r$ for C in the formula $A = \frac{1}{2}Cr$, we obtain the equation $A = \frac{1}{2} \times 2 \times \pi \times r \times r$, which after canceling and combining becomes the world famous $A = \pi r^2$. Exhausting argument but well worth it.

Archimedes wasn't a man who rested on his laurels. He then developed a clever algorithm to obtain a sequence of approximations to π, each one more accurate than its predecessor. If he had the time, he could have calculated this number to more places than a hand-held calculator displays. This is as good a place as any to debunk the myth that mathematicians always obtain "the answer." Often, all that is obtained is a procedure by which to obtain as accurate an answer as time and money permit. (Computer running time can be expensive.)

One could write an entire book on Archimedes' mathematical accomplishments. Since we must move on, let's summarize some of his achievements. He found the surface area and volume of a sphere of radius r to be $4\pi r^2$ and $\frac{4}{3}\pi r^3$, respectively. He found the volume of a cone of height h and radius r to be $\frac{1}{3}\pi r^2 h$.

Archimedes departed from the Greek penchant for theory and contemplation. He enjoyed building things and was willing to deal with inexact, even messy, quantities. He stands, in our minds, as one of the greatest mathematical geniuses of all time. It is with some reluctance that we must now bid him adieu.

Our next hero, Apollonius,[5] flourished in Alexandria circa 250 B.C. He studied the so-called *conic sections*: circles, ellipses, parabolas, and hyperbolas. This raises two pressing questions. What do these objects have in common? And what is "conic" about them?

As the "figures" in Figure 4-5 (so that's why we say figure! – that figures) illustrate, they are all obtained by intersecting a plane with a cone. Caution: The mathematician's cone is not to be found in an ice cream shop. The straight lines that make up the cone (they all meet at the vertex, by the way) extend to infinity in both directions, as the figure indicates. A horizontal plane cut forms a circle, while a tilting plane cut yields an ellipse. Many mathematicians will call

---

[5] Apollonius was interested in the problem of constructing a circle tangent to each of three given circles. This problem has been named after him.

us to task here – a circle is just a special kind of ellipse whose two axes of symmetry have the same length. Nevertheless, for heuristic purposes (look it up!), we present them separately.

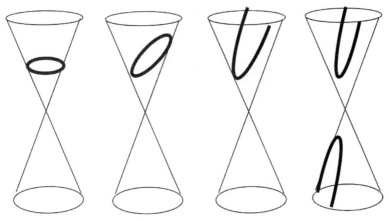

Figure 4-5

If a plane cut tilts enough, it becomes parallel to a line in the cone and never leaves the cone. This gives us a parabola. If we make the plane cut even steeper, we cut the bottom part of the cone, too, and form a hyperbola. Apollonius wrote all about these objects and discovered many of their remarkable properties. This was all done in the Greek spirit of contemplation of truth and beauty. Little did Apollonius know how central a role his conic sections would play almost 2000 years later. Our car headlights and our parabolic antennas hint at the many uses the parabola has. A thrown football moves in an upside-down parabolic path, as does, on a more sinister note, a bomb released from an airplane (not a smart bomb, but one that depends only on inertia and gravity). We may not like it, but warfare has often spurred humankind to great technological advances, like radar and jet planes – the latter a German, and the former a British invention in the dreaded Second World War.

Apollonius could not have foreseen the discovery of the Polish astronomer Kepler in the early seventeenth century (A.D.) that the earth follows an elliptic orbit around the sun. This fact escaped the attention of Copernicus who, though placing the sun at the center of the solar system, retained the "perfect" circular orbits of ancient cosmology. Apollonius' treatise "The Conic Sections" certainly traveled a great distance measured in centuries.

In the several hundred glorious years of Greek mathematics that we have been covering, a new nation was emerging on the Italian peninsula. First established in 509 B.C., the Roman republic grew in size and strength until it stretched all the way to the Po River by 268 B.C.. Ever at war with Carthage over dominance of the Mediterranean, Rome defeated that power in the third Punic War in 146 B.C. and established itself as a major world power. In that same year, Rome conquered Greece and much of Hellenistic Asia. In 27 B.C., Octavian changed his name to Augustus Caesar and established himself as Emperor of Rome. He put an end to the interminable wars of the past several centuries and established the Pax Romana – a two-hundred year period of peace and prosperity.

In the first century B.C., (perhaps it should be called "the last"), the great temporal power of Rome was felt throughout the Greek world. The Roman presence in Alexandria in no way diminished the cultural light of Greece. Greek mathematics flourished for several centuries, well into the A.D. period.

The Romans respected Greek culture and, in fact, brought many captured Greeks home to Rome to tutor their young. Roman writers often wrote in the style of the Greeks. Many mythological tales were rewritten in Latin. As to the conquered regions under their rule, as long as they paid tribute, they were left alone. The penalty for rebellion was unthinkable. So the Museum in Alexandria wagged on.

Sometime around 150 A.D., a mathematician/astronomer named Claudius Ptolemy[6] of Alexandria made a map of the ancient world in which he employed a coordinate system very similar to the latitude and longitude of today. Though the formal study of coordinates had to wait another 1500 years, we have in Ptolemy's map the seeds of this enormously important idea. We shall have to wait to read about it in this book – but less than 1500 years.

One of his most important achievements was his geometric calculation of *semichords*. Imagine a chord of a circle, that is, a line segment with both ends on the circumference of a circle. These endpoints determine an angle at the center of the circle, if we draw both radial lines, as in Figure 4-6.

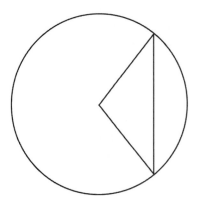

Figure 4-6

If we then add the line segment from the center to the midpoint of the chord, we get two congruent right triangles. Now in a right triangle, the sine of an acute angle, abbreviated $\sin \theta$ (where the Greek letter θ represents the angle) is defined as the ratio of the length of the opposite leg to the length of the hypotenuse. In Figure 4-7, under the assumption that the radius has unit length, that is, equals 1, we get that $\sin \theta$ equals the length of segment AM, which is exactly half the entire chord AB, hence the name semichord.

---

[6]Ptolemy obtained, using chords of a circle and an inscribed 360-gon, the approximation $\frac{377}{120}$ for π. Actually, he obtained the number $(3; 8, 30)_{60}$ which would have made the Babylonians proud.

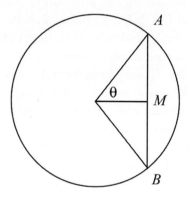

Figure 4-7

One may call Ptolemy the father of trigonometry, the study of triangles. The word "trigonometry" contains the Greek cognates for three, side, and measure. Put them together and you get the measurement of three-sided figures.

Having inherited the Babylonian and Greek view that Earth is the center of the universe, Ptolemy wondered why at certain times of the year it appears that Mars is moving backward. The technical term for this is *retrograde motion*. Since the circle is clearly the most perfect geometric figure, ancient Greeks expected the universe to conform to their aesthetic taste and make all orbits perfectly circular. So Ptolemy postulated a system wherein each body orbiting the earth spins on a circle of its own, called an *epicycle*. Figure 4-8 attempts to depict this.

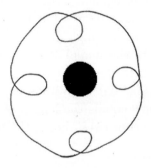

Figure 4-8

All told, Ptolemy required approximately 80 equations to describe, quite accurately, the locations of all the heavenly bodies of what we now call the solar system. Ptolemy's geocentric model went unchallenged until the middle of the sixteenth century and was defended by the Church for another few hundred years. A heliocentric (sun at the center) system first appeared in print in 1548 in a work written by Copernicus. Kepler and Galileo then advocated this theory in the early seventeenth century, though Kepler replaced his circles by ellipses.

Ptolemy's great works *The Almagest* and *Geographia* were translated by Islamic scholars and survived the ravages of history. We are indebted to these Islamic translators, without whom many manuscripts by the likes of Aristotle, Plato, and Euclid would have been lost.

1. Determine whether each of the following is prime or composite:

   (a) 37   (b) 29   (c) 330   (d) 4563
   (e) 45674   (f) 77236   (g) 101   (h) 61

2. Prove that a number is divisible by 2 whenever the digit in the
   "ones" place is even. (The even digits are 0, 2, 4, 6, and 8).
   How is this useful in Exercise 1?

3. Factor each of the following into primes:

   (a) 56   (b) 24   (c) 360
   (d) 450   (e) 3200   (f) 1000

4. After 2, why are all prime numbers odd?

5. Find the GCD of the following pairs of numbers:

   (a) 56, 72   (b) 24, 28   (c) 100, 360
   (d) 25, 450   (e) 150, 270   (f) 120, 300

6. The *least common multiple* (or LCM) of a pair of numbers a
   and b is the smallest number that is divisible by both a and
   b. It can be found by dividing the product of a and b by the
   GCD, that is, LCM $= \frac{a \times b}{\text{GCD}}$. Find the LCM of the following
   pairs of numbers:

   (a) 24, 35   (b) 24, 28
   (c) 20, 45   (d) 56, 72
   (e) 10, 24   (f) 25, 450

7. Show that all odd primes can either be written as $4n + 1$ or
   $4n + 3$, where n is a whole number. (The prime number 13, for
   example, is of the first kind, since $13 = 4 \times 3 + 1$; the prime
   number 19 is of the second kind, since $19 = 4 \times 4 + 3$.) Find ten
   primes in each category. Why is a number of the form $4n + 2$
   not a prime? What about a number of the form $4n$?

8. Use guesswork to find the GCD of the following pairs of numbers:

   (a) 20, 30   (b) 5, 85   (c) 18, 45
   (d) 40, 60   (e) 17, 19   (f) 1, 1000
   (g) 22, 33   (h) 24, 48   (i) 6, 60

9. Use Euclid's algorithm to find the GCD of the following pairs:

   (a) 15, 105   (b) 3, 36   (c) 90, 400
   (d) 400, 1100   (e) 355, 775   (f) 120, 300

10. Explain and prove the validity of the following criteria for con-
    gruence of triangles:

    (a) side, angle, side (abbreviated SAS)
    (b) angle, side, angle (abbreviated ASA)
    (c) side, side, side (abbreviated SSS).

11. Prove that opposite angles of a parallelogram are equal. Prove
    that the diagonals of a parallelogram bisect each other, that is,
    they intersect at their common midpoint.

12. In the accompanying figure, M is the midpoint of line segments
    $\overline{AB}$ and $\overline{CD}$. Prove that the two triangles are congruent. Why
    does it follow that $\angle C = \angle D$ and $\angle A = \angle B$? Now show that
    $\overline{AC}$ and $\overline{BD}$ are parallel.

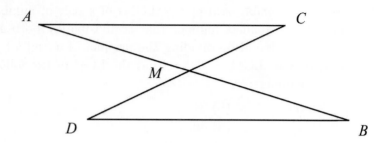

13. Prove that two right triangles are congruent if a leg and the hypotenuse of one triangle equal a leg and the hypotenuse of the other.

14. Using the approximate value of 3.14 for π, find the area of a circle with (a) a radius of 10 inches, and (b) a radius of 20 inches.

15. Show, using the formula $A = \pi r^2$, that doubling the radius of a circle multiplies the area by four. How does this explain the answers in the previous exercise? What is the effect of tripling the radius of a circle? Formulate a general rule yielding the effect on the area if we multiply the radius by any positive number. Use the formula for the volume of a sphere to predict a similar result if one doubles or triples the radius. (The formula is $V = \frac{4}{3}\pi r^3$.)

16. Study the streets and avenues of Manhattan on a map of New York City and compare that implicit coordinate system with latitude and longitude used in locating points on Earth. Then compare both systems with the coordinate system used to locate points in the xy-plane. You can see a map at the Web site given by the URL

   http://www.aaccessmaps.com/show/map/manhattan

Suggestions for Further Reading

1. Benson, Donald C. *The Moment of Proof: Mathematical Epiphanies.* Oxford University Press, New York, 1999.

2. Blatner, David. *The Joy of Pi.* Walker & Co, New York, 1997.

3. Dunham, William. *Journey through Genius: The Great Theorems of Mathematics.* Penguin, New York, 1991.

4. Gullberg, J., and Hilton, P. *Mathematics: From the Birth of Numbers.* W. W. Norton & Co., New York, 1997.

5. Stein, Sherman. *Archimedes: What Did He Do Besides Cry Eureka?* MAA, Washington, DC, 1999.

6. Struik, Dirk J. *A Concise History of Mathematics.* Dover, New York, 1987.

Chapter 5

Must All Good Things Come to an End?

The mighty Roman Empire did not result in the demise of Greek culture. In fact, Rome embraced the Greek worldview and admired the artistic and intellectual achievements of her vanquished neighbor. What, then, destroyed the energetic and exciting Greek way of life, characterized by a magnificent blend of mathematics, philosophy, art, architecture, literature, and so forth – neatly wrapped up in a package held together by the strings of a healthy, pagan, worldly love of life?

The classical worldview was not defeated by barbarians or by Roman legions. Ideas are too powerful to be defeated on the plains of battle. They can only be defeated by other ideas. The winners in an ideological battle need not be more correct or more logical. They simply must appeal to enough people willing to spread them – to fight for their acceptance.

The religious climate of the first century A.D. was mixed. On the one hand, there were the time-honored gods of Greece and Rome whose worship was linked to loyalty to family and state. There was, also, the thousand-year old monotheistic religion of the Jews, oddly enough respected by Rome. The Romans destroyed the temple for political reasons, not ideological ones. On the other hand, there were several recently established upstart religions – popular cults competing with the older, more established ways. These cults had several common characteristics. They promised immortality if the members believed in the cult figure. They had an initiation, or baptism, ceremony. Finally, they believed in an afterlife that was much more important than this one here on earth. Most of these cults, like the

Persian cult of Mithra or the Egyptian cult of Isis and Osiris are merely historical curiosities today. One of them has several billion followers today – Christianity, which is now anything but a cult.

The turning point in the spread of Christianity was its adoption by the Roman emperor Constantine as the official religion of Rome. (This is the same emperor who built the city of Constantinople – today Istanbul.) This assured its perpetuation after Rome fell in the fifth century (A.D.). By the year 1100, Christianity was the religion of Europe with the exception of Moslem-occupied southern Spain.

It was the otherworldly philosophy of early Christianity that spelled the end of classical thought. After all, if this world is just a testing ground for admission to heaven where the immortal soul will reside for eternity, then why make a fuss about this world? Moreover, the world is full of corruption, sin, and falsehood – certainly not a worthy place for painting or sculpting. Truth, of course, is not reached through observation or Socratic dialogue, but rather from revelation. The Bible says it all.

Even the Greco-Roman interest in astronomy/astrology is absurd. An omnipresent god who may be invoked through prayer replaced the mythological gods residing on the planets. Why pursue the study of mathematics when the real issue is salvation? This view, together with the barbarian invasions and the collapse of Rome, paved the way for the Dark Ages. The lights went out in Europe and would come back only dimly in the Middle Ages, as we shall see.

In Alexandria, one of the last holdouts of classical Greek thought, extant as late as the fifth century, a Christian mob set fire to the library and brutally killed Hypatia, a pagan mathematician (and one of the few famous women of antiquity in mathematics).

In the century before the fall of Rome, there occurred a division in the Roman Empire with the establishment of the Eastern Empire whose capital was Constantinople. Later called the Byzantine Empire, it lasted until 1453 when it was overrun by the Seljuk Turks. In the last century of the Eastern Empire's existence, Byzantine artists and intellectuals fled to Renaissance Italy and brought with them a large number of ancient Greek manuscripts, thus reintroducing the influence of the classical world, thereby serving as a catalyst of the Renaissance (the rebirth).

To some, history is viewed as a succession of regimes. It is the repetitive play of "The Rise and Fall of ...", replete with a host of dates, that characterizes, to some, the march of events through the ages. While this view has some truth to it, it is the march of ideas through time that constitutes a significant part of history. And ideas can perish! Books, unfortunately, can be burned or lost. Proponents of an idea can be slain or jailed. The romantic poet wants to shout, "but the ideas live on!" while the realist accepts that ideas can be suppressed, at least for a while. Intolerant regimes, since the dawn of recorded history, have retarded or even eliminated movements through evil repressive measures.

In light of all this ranting, what a breath of fresh air awaits us as we study the Golden Age of Islam. Rising out of the desert in the seventh century on the coattails of Judaism and Christianity, a new monotheistic religion, Islam, united the various people of the Middle East with a mission of spreading the submission to the will of Allah (the Muslim name for God). Allah spoke through his messenger Mohammed who recorded his words in the Koran, the Muslim equivalent of the Bible. Having its origins in 622 with the flight of Mohammed from Mecca to Medina (known as the Hegira, meaning flight), the new religion spread like wildfire. By the year of Mohammed's death (632), the entire Arabian Peninsula had " submitted" to the will of Allah, perhaps influenced by the armed hordes of fanatical believers who made offers that couldn't be refused.

At any rate, the mission of Mohammed and his followers was wildly successful (the sword is sometimes mightier than the pen) and by the ninth century, the great caliphs seated in Baghdad held sway over an empire stretching from Morocco to Tibet – the largest empire up to that time.

Islamic designs on Europe heralded by an invasion of Spain in 711 were decisively stopped by the army of Charles Martel[1] (688-741) in France at the Battle of Tours which ended in 732. (His grandson was Charlemagne, or Charles the Great). At about this time, Islamic incursions into Europe through Constantinople were thwarted by the army of Byzantium under Leo III (680-741). Christian control of Europe was now assured.

The army of the Caliph Omar captured Alexandria in 641. When asked by his general what to do with the great library there (the very same one where Euclid had taught almost a thousand years before this date), Omar is alleged to have replied that if the teachings agreed with the Koran, the works were unnecessary and the library

---

[1]He was known as "the Hammer."

should be burned to the ground. On the other hand, if the teachings contradicted the Koran, the works were evil, and the library should – you guessed it – be burned to the ground.[2]

Perhaps movements mellow with time. About one hundred years later, the great caliph Harun Al Rashid (765-809) and his successor Al Mamun (786-833) were patrons of the arts and sciences. The new capital city of Baghdad, the construction of which started in 762, became the cultural center of the enormous caliphate. Al Mamun invited scholars and artists to his newly built "House of Wisdom" – a university in which contributors of all faiths were tolerated – and there flourished an outpouring of astronomy, mathematics, medicine, chemistry, literature, and so forth. In addition, much translating of Greek classics went on, thereby helping to preserve the works of great scholars such as Aristotle, Euclid, and Ptolemy for later generations.

---

[2] As a friend once said, "Death to fanatics!"

Mathematics from India, such as the influential work of the great Brahmagupta[3] (598-665), introduced several interesting ideas, some of which, along with the work of Arabic mathematicians, developed into the Hindu-Arabic decimal number system we use today. They used a symbol for zero and had speedy algorithms to add, subtract, multiply, and divide with their numbers. Using only ten digits, they could represent any number using, of course, the brilliant position system. An idea first utilized by the Babylonians reappeared in a much-improved version.

In stark contrast to the Islamic mathematicians and merchants, their European counterparts were using clumsy Roman numerals still seen today on huge public clock towers. In case you have forgotten, Roman numerals are summarized in the box below.

$$I = 1$$
$$II = 2$$
$$III = 3$$
$$IV = 4$$
$$V = 5$$
$$VI = 6$$
$$VII = 7$$
$$VIII = 8$$
$$IX = 9$$
$$X = 10$$
$$XX = 20$$
$$L = 50$$
$$C = 100$$
$$D = 500$$
$$M = 1000$$

ROMAN NUMERALS

Notice how the complicated looking number

MMMDCCCLXXVIII

---

[3]He had approximated the length of the year to be 365 days, 6 hours, 5 minutes, and 19 seconds.

is simply 3,878 in the Hindu-Arabic system. Now imagine the problem $3,878 \div 364$ being done by a Roman accountant! If we were in a sarcastic state of mind, we would wonder whether this contributed to the decline of Rome.

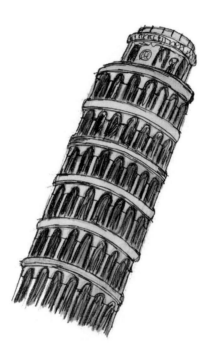

These clever Hindu-Arabic numerals finally found their way into Europe not only through books but, in addition, through trade between Christian and Arabic merchants in the ports of Italy. Arabic merchants brought oriental goods (with hefty markups!) to European ports hungry for silk and other Chinese goods. They calculated very quickly, to the amazement of their sluggish European customers. One such merchant, Leonardo of Pisa, also called Fibonacci, was a mathematician in his spare time. He wrote a book *Liber Abaci* in approximately 1200 that popularized the new Arabic arithmetic. Of course, this was before the printing press, and there certainly was no Manuscript-of-the-Month Club. Nevertheless, several handwritten copies spread across Europe and attracted some attention. The system met initial resistance, and one Italian community even outlawed their use, citing as the reason, that the copies were easy to alter.

A speedy method for multiplication that was introduced around this time is known as *lattice multiplication* or *Hindu lattice multiplication*. This method is demonstrated in the example below.

Example 5-1 *Multiply* 276 × 49 *using lattice multiplication.*

> The first step is to draw a 3 x 2 grid (lattice) like the one below. (It is 3 x 2 because the two factors have 3 and 2 digits, respectively.)

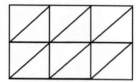

> Now, write 276 along the top and 49 along the right side.

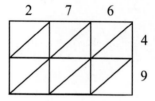

> The next step is to fill in the lattice by multiplying the two digits found at the head of the column and to the right of the row. When these intermediate products result in a two-digit number, the tens digit is placed above the diagonal and the ones digit is placed below. If the intermediate product is a single-digit number, a 0 is placed above the diagonal. Thus, we have

Now to get the final product, we add along the diagonals
starting at the lower right. Any "carries" that result from
adding are carried to the next diagonal as we have

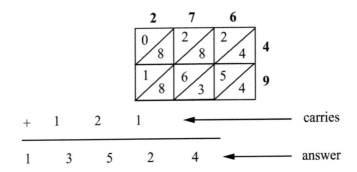

+   1   2   1   ←————————— carries

1   3   5   2   4   ←———————— answer

and the answer is 13,524.   ∎

Another vehicle in which Arabic mathematics rode into Europe
was a book written by the great Islamic mathematician Mohammed
Ibn Musa Al Khwarizmi[4] (780-850), *Hisab al-jabr w'al-muqabala* –
the science of transposition and cancellation. It is from *al-jabr* that
we get the term *algebra*. While this book handles linear and quadratic
equations skillfully, negative numbers do not yet make an appear-
ance. Hundreds of years later, European mathematicians would still
treat them as absurd quantities.

Al Khwarizmi wrote another very influential book, of which only
a Latin translation survives, in which he outlined the Hindu position
system, including the use of zero, and the various procedures used
to perform the basic arithmetic operations. In translation, his work
became *Liber Algorismi* (The Book of Al-Khwarizmi), from which
algorithm evolved.

Algebra is not the only technical term that comes from Arabic. So
do many other technical terms such as azimuth and cipher. Islamic
mathematicians did much algebra and streamlined the execution of
its manipulations. Instead of solving $x + 5 = 20$ by subtracting
5 from both sides, they simply transposed the +5 on the left side
of the equation to the right, where it reappears as a −5, yielding

---

[4]His mathematics was done entirely in words, using no symbols.

$x = 20 - 5$, which is the correct answer, 15. Similarly, the equation $3x = 12$ becomes, after transposing the multiplier 3 to the other side $x = \frac{1}{3} \times 12$, again yielding the correct answer 4. Notice that the multiplier 3 became the "divide by 3" operation, written here as multiplication by $\frac{1}{3}$. But beware, warned the Arab algebraists! Remember the order of operations and reverse them on the other side.

Thus, the equation $3x + 5 = 35$ involves transposing the $+5$ and then transposing the 3 multiplier. We get $3x = 35 - 5 = 30$ and then $x = \frac{1}{3} \times 30 = 10$. In one step, this could be written $x = \frac{1}{3} \times (35 - 5) = \frac{1}{3} \times 30 = 10$. The parentheses remind us to subtract the 5 before multiplying by $\frac{1}{3}$.

This can be understood using the "shoes and socks" argument. In the morning you put on your socks and then your shoes. In the evening you reverse the process (we hope) and take off your shoes and then your socks. In the equation $3x + 5 = 35$, x gets up and multiplies itself by three and then adds the 5. To undo this and recover the value of x, we must first subtract 5 and then divide by 3 (or equivalently multiply by $\frac{1}{3}$). How about a big hand for these clever algebraists?

The Persian mathematician/poet Omar Khayyam[5] (1040-1120) boldly went to the cubic equation with partial success, extending the work of his Islamic predecessors such as Al-Khwarizmi. He continued the trend in classical Greek mathematics of searching for geometric solutions to algebraic problems – a trend that influenced a host of later-day European mathematicians, such as Descartes. Omar Khayyam's great poem *Rubaiyat* was translated into English and reveals his thinly disguised secular worldview, quite uncommon in his time.

On a sad note, the splendid Golden Age of Islam came to a horrible end in the thirteenth-century when Mongol invaders sacked the Middle East. Genghis Khan destroyed Baghdad in 1258, and in the words of a witnessing historian, "rivers of blood" flowed down the streets of that glorious city. The Mongols, incidentally, had a two-hundred year run of terror which reached as far west as Krakow, Poland. They were among the most merciless, brutal warriors ever

---

[5]Khayyam wrote three books, one on arithmetic, one on music, and one on algebra, all before he was 25 years old. He also measured the length of the year to be 365.24219858156 days which is extremely accurate for back then.

known and would have rivaled the slaughter and destruction wreaked by the unspeakably evil Nazi Germany but for their lack of twentieth-century weapons. Fortunately, troubles back home resulted in their retreat. We still speak of hordes – large military units in their armies (akin to our divisions or regiments).

In summation, Islamic mathematicians are remembered for three important activities:

1. Improvements in algebra, the science of manipulating unknown quantities,

2. Hindu-Arabic numerals and their associated algorithms, and

3. Translation of untold numbers of ancient Greek manuscripts that might otherwise have been lost forever.

Prejudiced Westerners of today would do well to remember the age during which Islamic culture was glorious while European culture was practically nonexistent. While we are on this topic, Greek mathematics comes from Egypt and Babylon – one in Africa and the other in Asia – and our numerals originated in India. Moreover, though it didn't feed into European universities of the past millennium, many cultures, such as the Mayans and the Chinese, produced interesting and clever mathematics. In the modern age, the world mathematical community has productive members in every country. Enough said.

Exercises

1. Write each of the following in Roman numerals:

   (a) 145   (b) 237   (c) 1,865
   (d) 1,492 (e) 2,189 (f) 3,999

2. Write each of the following in our modern notation:

   (a) MMCCLIX   (b) MMCMLVII   (c) DCCCLIV
   (d) CCXLIX   (e) MDCXXXIII   (f) MMMDCLXXVIII

3. Try to add the Roman numerals of Exercise 2 and develop an algorithm for adding Roman numbers.

4. Develop a shortcut for multiplying modern numbers by 10, by 100, by 1000, by any power of 10. Now do this for division by any power of 10. Justify your algorithm.

5. Explain why division by a whole number n is the same as multiplication by $\frac{1}{n}$. For example, division by 5 is the same as multiplication by $\frac{1}{5}$.

6. Why can Exercise 5 apply even when n is not a whole number? In other words, why is division by $\frac{a}{b}$ the same as multiplication

by $\frac{b}{a}$? In symbols, why is this equation true?

$$x \div \frac{a}{b} = x \times \frac{b}{a}$$

7. Multiply each of the following using the lattice method:

   (a) 43×67
   (b) 39×46
   (c) 478×42
   (d) 189×37
   (e) 214×288
   (f) 527×747
   (g) 1243×567
   (h) 12345×4321

8. There is much more to the mathematics of India then we have mentioned here. Research this topic and summarize your results in a one-page paper.

9. Traditional Chinese (approximately 1500 B.C.) used a *multiplicative number system* where symbols from 1 to 9 are used to multiply the appropriate base-10 symbol. For example, $2,465$ would have been written as

since they wrote their numbers vertically (see the chart on the next page). Write the following numbers in Traditional Chinese:

   (a) 678   (b) 5,812   (c) 98,761
   (d) 12,345   (e) 23,789   (f) 2,112

NUMBER	SYMBOL
1	一
2	二
3	三
4	四
5	五
6	六
7	七
8	八
9	九
10	十
100	百
1000	千
10,000	万

CHINESE NUMBERS

10. Write each traditional Chinese numeral as a Hindu-Arabic numeral.

a. 四
   千

   三
   百
   五
   十
   五

b. 九
   百
   七
   十
   六

c. 八
   千

   六
   百
   九
   十
   二

11. The Mayans also wrote numbers vertically. They wrote their numbers with sticks and stones in base-20, or the *vigesimal* system. Research the Mayan system and write a one-page paper summarizing your results.

12. Brahmagupta's formula gives the area A of a *cyclic quadrilateral*, that is, a simple quadrilateral that is inscribed in a circle, with sides whose lengths are a, b, c, and d. The formula is

$$A = \sqrt{(s-a)(s-b)(s-c)(s-d)}$$

where s is the semiperimeter given by $\frac{1}{2}(a+b+c+d)$. Use this formula to find the area of a cyclic quadrilateral with sides $a = 2, b = 6, c = 6$, and $d = 8$.

13. Develop a proof for Brahmagupta's formula given in Exercise 12.

14. Al Khwarizmi used no symbols, like x, in his algebra. Obtain the answer to this question mentally by using his *method of inversion*. "Subtract three from a certain number. Then multiply by two. Square this result and you have 16. What is the number?"

15. In the decimal number 0.23, the 2 represents tenths while the 3 represents hundredths, that is, $0.23 = \frac{2}{10} + \frac{3}{100}$. On the other hand, you are no doubt aware that $0.23 = 23/100$. Reconcile these two expressions for 0.23 and, while we are on the subject, explain why $0.23 = 23\%$. Why is it that a decimal may be

converted into a percentage by shifting the decimal point two places to the right? Write these decimals as fractions and as percents:

(a) 0.25   (b) 0.5   (c) 0.75   (d) 0.3
(e) 0.05   (f) 0.125   (g) 1.5   (h) 2
(i) 0.001   (j) 1.25   (k) 10   (l) 0.333...

Suggestions for Further Reading

1. Bell, E. T. *Men of Mathematics*. Touchstone Books, New York 1986.

2. Datta, B. *Ancient Hindu Geometry: The Science of the Sulba*. South Asia Books, Bombay, 1993.

3. Dzielska, Maria. *Hypatia of Alexandria, Revealing Antiquity, No. VIII*. Harvard University Press, Cambridge, MA, 1996.

4. Edeen, S., and Edeen, J. *Women Mathematicians*. Dale Seymour Publications, Parsippany, NJ, 1990.

5. Gristien, Louise S., and Campbell, Paul J. *Women of Mathematics: A Biobibliographic Sourcebook*. Greenwood Press, New York, 1987.

6. Osen, Lynn M. *Women in Mathematics*. MIT Press, Boston, 1975.

7. Perl, Teri. *Women and Numbers*. Wide World Publishing/Tetra, San Carlos, CA, 1993.

Suggestions for Further Reading

Chapter 6

Europe Smells the Coffee

Near the end of the (Western) Roman Empire, the breakdown in trade brought about a new way of rural life – the self-sufficient manor. This was a herald of things to come. In fact, the manor was exactly what the Dark Ages called for. The wide and well-built Roman roads (not all of which led to Rome) were hardly traveled with the exception of a few pilgrims on their way to holy sites. There were no universities and there was little need for mathematics or, for that matter, any intellectual endeavor. The priests were generally the only ones who were literate and the only justification for any mathematical activities on their part was the calculation of the date of Easter, which unlike Christmas, depends on the lunar calendar.

The average peasant lived and died within fifty miles of his birthplace. Under the new order, feudalism, he was a slave to his lord and worked the latter's land in return for a tiny portion of it for sustenance. With several minor exceptions, there was little or no scholarship throughout Europe, though monks in the monasteries wrote and rewrote ancient manuscripts and studied the Bible and the works of Roman authors such as Boethius (480-524). Very few people could read, let alone write. The priests read Bible passages to them and the church provided the rituals, holidays, and all of the other components of the social structure of that age.

The monks sang monophonic, free-flowing chants with haunting melodies of Pope Gregory (540-604) – the *Gregorian chants*, of course! – year after year, decade after decade. Life was stagnant. Virtue consisted of self-denial, prayer, and toil. The idea of a Greco-Roman bathhouse where men of letters congregated and debated philosophy was as distant as the outermost galaxies of the universe. To be sure, there was a brief period of learning in the so-called Carolingian Renaissance during the rule of Charlemagne, but it doesn't amount to a hill of beans.

The political agenda of the rulers of Europe was consolidation of power, repelling the Muslim invaders, and conversion of the barbarians to Christianity. This conversion effort spanned the five hundred years of the Dark Ages. There were no significant advances in mathematics in Europe during this time.

What changed? There are volumes written about this question. It was the common belief that Christ would return to earth in the year 1000. As you probably are aware, he didn't. It was therefore logical to assume that civilization would endure another thousand years, as humankind had not yet acquired the terrifying nuclear weapons of the twentieth century. What followed was a church-building frenzy. This required a bit of carpentry, stonemasonry, transportation of goods from distant communities, and the hiring of laborers and artists. Feudal lords discovered that it was more efficient to buy armies than to extract servitude from their vassals. It was better to rent land to the peasants in return for much needed cash. This was the death knell of feudalism. Gradually a money economy began to take shape. Many restless young men ran off to the cities to join merchant guilds or join the army – talk about upward mobility! The first crusade of 1095 woke Europe up and made it aware of a larger world. A greater demand for silk and other oriental products made the merchants of the Italian city-states prosper and these middle-class communities acquired power and influence, in some cases even self-rule. The twelfth century saw the rise of several European universities, such as Oxford, Paris, and Bologna, with many more soon to follow.

It was in the twelfth century that translations of works by Aristotle, Euclid, and other great Greek writers started to appear. This glimpse into the past glories of the classical world excited the imaginations of the Europeans.

The High Middle Ages was a time of growth and awakening. Huge gothic cathedrals such as Notre Dame and Chartres were monuments to the scope and vigor of activity in that period. New ideas appeared in Europe such as gunpowder, spectacles, mechanical clocks, and the flying buttress that permitted gothic churches to have large stained glass windows by freeing the walls from having to support the roof.

New mathematical ideas and the new Hindu-Arabic numerals gave scholars a shot in the arm. Commercial arithmetic was suddenly in demand and by the thirteenth century, merchants in Italy established private schools for their future heirs. At approximately this time, the Hanseatic League, an alliance of German, Swedish, Danish and English towns negotiated by their guilds to fight piracy and promote commerce, established schools that required commercial mathematics.

While the twelfth century saw a heightened interest in mathematics and learning, much of it was scholastic in nature. It was the epoch from which we draw the sarcastic question, "How many angels can sit on the head of a pin?" The mathematics was astonishingly simple, as were the other subjects. The texts were handwritten versions of ancient authors. There was, of course, no empirical science and the highest authority in any matter was the Bible. To be sure, a battle was brewing between faith and reason – between revelation and observation – between authority and free inquiry.

Enter Thomas Aquinas (1225-1272) and his great peace making compromise. Reason, he said, is valid in matters of this world, while revelation is valid in matters of the next one. The word of God is absolute and final on matters such as birth, death, prayer, salvation, duty, and so forth, while one need not consult the Bible to find out how to boil water, hunt boar, or to solve a quadratic equation. Freedom at last! Use reason to your heart's content, but . . . don't dare interfere with Church doctrine.

This truce lasted for several hundred years but, as we shall see later, was doomed from the start. In the interim, the teaching of the great philosopher Aristotle, to cite an example, was banned by the Church and hence removed from the curriculum of the University of Paris. The mathematics of Euclid wasn't deemed to be a threat to dogma and was spared the censor's cut. Little did the defenders of the faith realize that their greatest challenge would come from that seemingly harmless subject. We will see how this occurred when we get to the sixteenth century.

This is as good a time as any to look at the mathematics of Fibonacci.[1] He listed the following sequence of numbers: 1, 1, 2, 3, 5, 8, 13, 21, 34, 55, 89, 144, ..., where the ... means keep on going until we tell you to stop. In this case, the sequence continues forever. The trick is to see the pattern so that you can continue the sequence. The pattern is simple. Each term is the sum of the two preceding terms. Thus, 13 appears where it does because it is the sum of the two preceding terms 5 and 8. The third term of the sequence, 2, is the sum of the two preceding terms. This pattern does not explain the first two terms. They are simply given to us by Fibonacci.

Now each term can be denoted using subscripts that identify the order in which the terms appear. We denote the Fibonacci numbers by $u_1, u_2, u_3, \ldots$, and the nth term is denoted u_n. The sequence may now be presented by stating the first and second terms and the

---

[1] Fibonacci's sequence of numbers occurs in many places including Pascal's triangle, the binomial formula, probability, the golden ratio, the golden rectangle, plants and nature, and on the piano keyboard, where one octave contains 2 black keys in one group, 3 black keys in another, 5 black keys all together, 8 white keys, and 13 keys in total.

recursive relation which says, in an equation, that each term is the sum of its two predecessors.

$$u_1 = 1$$
$$u_2 = 2$$
$$u_{n+2} = u_n + u_{n+1}$$

The last equation in the box says that the $(n+2)$nd term is the sum of the nth term and the $(n+1)$st term. When $n = 1$ for example, this says that the third term, u_3, is the sum of the first term, u_1, and the second term, u_2, which is, of course, correct. Sequences play an important role in mathematics today, and it's interesting to see that the concept is quite old.

It should be obvious to you that the sequence terms grow large fairly rapidly, eventually breaking through any arbitrarily set "ceiling." If a macho mathematician were to lay down the challenge, "Will your terms ever exceed 1,000,000,000,000?" we would respond, with confidence, "Absolutely!" After some calculating, we would, to the accompaniment of a drum-roll, present a Fibonacci number larger than a trillion. Another way of saying all this is that the Fibonacci numbers approach infinity.

On the other hand, consider the sequence of ratios $\frac{1}{1}$, $\frac{2}{1}$, $\frac{3}{2}$, $\frac{5}{3}$, $\frac{8}{5}$, $\frac{13}{8}$, ... , formed by dividing each term by its predecessor. Instead of consistently getting larger, they alternate between growing and shrinking! In decimal form, the first few ratios are 1, 2, 1.5, 1.66$\bar{6}$, 1.6, 1.625, ... Since this alternation is consistent, successive numbers narrow the range in which future ratios can fluctuate. It turns out that the ratios approach a single target which we can readily calculate using a clever argument.

Let us call the target (or *limit* as mathematicians say) L. Furthermore, let us denote the nth ratio R_n. In other words (symbols?), $R_n = \frac{u_{n+1}}{u_n}$. Observe that if we divide the last equation in the box (the recursive relation) by u_{n+1}, we get

$$\frac{u_{n+2}}{u_{n+1}} = \frac{u_n}{u_{n+1}} + \frac{u_{n+1}}{u_{n+1}}$$

which, after applying the new symbols for the ratios, becomes

$$R_{n+1} = \frac{1}{R_n} + 1$$

Notice that the first ratio on the right side of the equation is the reciprocal of the ratio we have defined. It is the ratio of the nth term over its successor – the $(n+1)$st term. Now let n approach infinity, and replace both ratios by their limiting value L, and we get

$$L = \frac{1}{L} + 1$$

which, after courageously multiplying both sides by L, becomes

$$L^2 = 1 + L$$

Finally, we transpose everything to the left side, obtaining the quadratic equation

$$L^2 - L - 1 = 0$$

Recall the quadratic formula, which we now invoke to solve this equation. The general form of a quadratic equation is $ax^2 + bx + c = 0$, in which a, b, and c are the given coefficients which distinguish one quadratic equation from another. In the equation, $5x^2 + 2x - 1 = 0$, for example, $a = 5$, $b = 2$, and $c = -1$. The quadratic formula gives us solutions in terms of a, b, and c. It says that

$$x = \frac{-b \pm \sqrt{b^2 - 4ac}}{2a}$$

In our quadratic equation, $L^2 - L - 1 = 0$, we see that $a = 1$, $b = -1$, and $c = -1$. Choosing the plus sign in front of the square root symbol yields the solution $L = \frac{1+\sqrt{5}}{2}$, which is approximately 1.618. (Choosing the minus sign in front of the square root symbol would have given us an absurd, negative answer.)

We have obtained a celebrated number dear to all mathematicians and to many artists. It is called the *golden ratio* and was known to the Ancient Greeks as the most pleasing ratio of the length of a rectangular painting frame to its width. Notice that very few paintings are square, by the way, and even more rare is it to see a painting on a canvas which is twice as long as it is wide.

Here is how they reasoned. Using the notation of today, call the width one and the length L. There is no loss of generality in calling the width one, since we are only after the ratio of length to width. Thus a 10-by-15 rectangle has the same ratio as a rectangle with dimensions 1 by 1.5. Now they assumed that the perfect or *golden* rectangle has the property that the removal of a square from it leaves a (smaller) rectangle that is similar to the original one. Figure 6-1 illustrates this idea. AD has length one, and DF has length L.

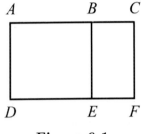

Figure 6-1

Since $ABED$ is a square, the length of DE is one and it follows by subtraction that the length of EF is $L-1$. Now the dimensions of the rectangle $BCFE$ are $L-1$ by 1, while the dimensions of the original rectangle are 1 by L. They will be similar if they have the same ratio of length to width, yielding the equation $\frac{L}{1} = \frac{1}{L-1}$.

If we cross multiply, we obtain $L(L-1) = 1$. Distribute the L on the left and transpose the 1 on the right, and we have $L^2 - L - 1 = 0$. Wow!! This is the quadratic equation of the Fibonacci ratios. Then it must have the same solution, namely $L = \frac{1+\sqrt{5}}{2}$. The face of the Parthenon in Athens has been seen as a golden rectangle and so have many other facades in Greek and Renaissance architecture. The golden ratio appears in many strange places in both the natural world and the human world of magnificent artistic and scientific achievements. Psychologists have shown that the golden ratio subconsciously affects many of our choices, such as where to sit as we enter a large auditorium, where to stand on a stage when we address an audience, and so on, and so on.

Let's look at some simple examples.

Example 6-1 *Solve $x^2 - 5x - 6 = 0$ by using the quadratic formula.*

The equation is in standard form $ax^2 + bx + c = 0$ with $a = 1$, $b = -5$, and $c = -6$. Substituting these into the quadratic formula we have

$$x = \frac{-(-5) \pm \sqrt{(-5)^2 - 4(1)(-6)}}{2(1)}$$

$$= \frac{5 \pm \sqrt{25 + 24}}{2}$$

$$= \frac{5 \pm \sqrt{49}}{2}$$

$$= \frac{5 \pm 7}{2}$$

The $\pm$ means there are two answers. The one with the $+$ is $\frac{5+7}{2} = \frac{12}{2} = 6$ and the one with the $-$ is $\frac{5-7}{2} = \frac{-2}{2} = -1$. ∎

Example 6-2 *Generate the first five terms of the sequence given by the recurrence relation* $u_{n+2} = 2 \times u_{n+1} + u_n$, *with the initial conditions* $u_1 = 1$ *and* $u_2 = 1$.

The recurrence equation says that one gets any term by doubling the preceding term and adding the one before that. Therefore, the third term is given by the formula $u_3 = 2 \times u_2 + u_1$ or $u_3 = 2 \times 1 + 1 = 2 + 1 = 3$. The fourth term is given by the equation $u_4 = 2 \times u_3 + u_2$ or $u_4 = 2 \times 3 + 1 = 6 + 1 = 7$. Finally, we have $u_5 = 2 \times u_4 + u_3$. Substituting the values $u_3 = 3$ and $u_4 = 7$, we have $u_5 = 2 \times 7 + 3 = 14 + 3 = 17$. ∎

Returning to the Fibonacci numbers, suppose you are a schoolteacher and you have n children whom you are lining up for some activity. You do not wish to have two boys next to each other on the assumption that boys are more likely to fight with one another. In how many ways can this be done? Let's try this with $n = 1$. Clearly, there are two solutions which we will represent by the arrays B and G, for one boy and one girl. Now when $n = 2$, we have three possibilities, namely BG, GG, and GB. (BB is out because it entails two adjacent boys – who are probably fighting as we speak!) When $n = 3$, we have the five solutions BGG, GGG, GBG, BGB, and GGB. Let's do one more case and then make a guess. When $n = 4$, the eight solutions are BGGG, GGGG, GBGG, BGBG, GGBG, BGGB, GGGB, and GBGB. Let's summarize these findings.

n	Number of ways
1	2
2	3
3	5
4	8

It does appear that we get the Fibonacci numbers – but four cases do not constitute a proof. In fact, neither do four million cases! In light of the fact that we are dealing with infinity, what are four million cases? We need to show that the recursive relation governing the Fibonacci numbers works here, too.

To this end, let us divide all lines of $n + 2$ boys and girls into two groups: those that end with a boy and those that end with a girl. In the first group, the second to the last child must be a girl since two boys cannot be next another, remember, that is, all lines in the first group look like this: ...... GB. The second group has no restriction on the gender of the next to the last child. Then the first group has the same size as the number of lines of n children, while the second group has the same size as the number of lines of $n + 1$ children. But this is the recursive relation of the Fibonacci numbers!

Before we discuss the disappointments of the fourteenth century in detail, let's start with something optimistic. Several new universities were founded and commerce continued to mushroom. The Italian city-states prospered and many merchants became patrons of the arts, causing the migration of musicians, painters, architects, and so forth, to those communities. Ancient Greek manuscripts from Byzantium and Arabic translations of ancient works captured at Toledo in 1085 increased the cultural influence of the classical world on the intellectuals and artists of the times, resulting in the great Renaissance. This rebirth involved a heightened interest in this world and curiosity about the art, mathematics, and philosophy of the pagan classical culture.

Nicole Oresme,[2] a French priest and mathematician, was curious about *infinite series*, that is, sums of infinitely many terms. He read about the ancient Greek discovery that the sum of the reciprocals of the numbers 1, 2, 4, 8, 16, 32, ... (the so-called powers of 2) was 2. This astonishing fact can be seen as follows. Consider the sum $1 + \frac{1}{2}$. This is $\frac{3}{2}$ and it is less than 2. In fact, $\frac{3}{2} = 2 - \frac{1}{2}$. Now consider the sum $1 + \frac{1}{2} + \frac{1}{4}$. This is $\frac{7}{4}$, or $2 - \frac{1}{4}$. The next *partial sum*

---

[2] Oresme was the first to prove the harmonic series diverges.

$1 + \frac{1}{2} + \frac{1}{4} + \frac{1}{8} = \frac{15}{8} = 2 - \frac{1}{8}$. We have a clear pattern. The sum of the terms up to $\frac{1}{2^n}$ is $2 - \frac{1}{2^n}$. Now as n goes to infinity, so does 2^n. Then its reciprocal $\frac{1}{2^n}$ goes to zero. Then the partial sums clearly go to 2. This is a modern argument – not the ancient one. It is amazing that infinitely many terms can add up to a finite sum! In mathematics, this is known as *convergence*.

At any rate, Oresme wondered about the sum of the reciprocals of all the integers, that is, the sum

$$1 + \frac{1}{2} + \frac{1}{3} + \cdots + \frac{1}{n} + \cdots$$

and proved that this sum is not as lucky as the sum of the reciprocals of powers of 2. This sum increases without bound and goes to infinity. It is said to *diverge* to infinity. To see this, consider the groups

$$\frac{1}{2}$$

$$\frac{1}{3} + \frac{1}{4}$$

$$\frac{1}{5} + \frac{1}{6} + \frac{1}{7} + \frac{1}{8}$$

$$\frac{1}{9} + \frac{1}{10} + \frac{1}{11} + \frac{1}{12} + \frac{1}{13} + \frac{1}{14} + \frac{1}{15} + \frac{1}{16}$$

$$\frac{1}{17} + \frac{1}{18} + \frac{1}{19} + \frac{1}{20} + \cdots + \frac{1}{29} + \frac{1}{30} + \frac{1}{31} + \frac{1}{32}$$

where the "$\cdots$" tells us to supply the missing terms. A careful count will reveal that these groups have 1, 2, 4, 8, 16, and 32 terms, respectively. Furthermore, each group is larger than (or equal to, in the first case) $\frac{1}{2}$. The third group, for example, consists of four terms, each larger than $\frac{1}{8}$. The sum is, therefore, larger than $\frac{4}{8}$ which equals $\frac{1}{2}$. Since we can continue this grouping process forever, we can amass as many $\frac{1}{2}$'s as we wish. Alas, the sum, known as the *harmonic series*, must be infinite.

Infinite series became extremely important in the seventeenth and eighteenth centuries when they enabled mathematicians to compute values of logarithms, roots, and trigonometric functions to any desired degree of accuracy.

The seventeenth-century German mathematician and philosopher, Leibniz[3] (1646-1716) discovered the following series for π:

---

[3]Leibniz is considered to be one of the fathers of calculus.

$$\pi = 4 \times \left(1 - \frac{1}{3} + \frac{1}{5} - \frac{1}{7} + \frac{1}{9} - \frac{1}{11} + \cdots \right)$$

It is a mystery that the ratio of the circumference to the diameter is four times the alternating sum of the reciprocals of all the odd numbers! This is but one of the many mysteries surrounding π.

Many never-ending decimals yield clever sums as the following trick shows. Consider the number $0.5555555\ldots$. Let us call it N. So we have

$$N = 0.55555555\ldots$$

Then multiplication by 10 yields

$$10N = 5.55555555\ldots$$

It follows by subtraction that $9N = 5$, since the decimal parts of the previous two equations cancel when we subtract. Then we get $N = \frac{5}{9}$. Now the digit 5 has no particular significance. We have the exciting fact that the *nonterminating* decimal $0.aaaaaaaaaaa\ldots$ is just $\frac{a}{9}$. The a represents any single digit. It should be mentioned that this shows that when $a = 9$, we get the equation $0.99999\ldots = 1$.

What about a repeating decimal like $0.474747474747\ldots$? Again, let's call it N. Then

$$N = 0.4747474747474747\ldots$$

Multiplying both sides by 100 yields

$$100N = 47.47474747474747\ldots$$

It follows, once again, by subtraction and cancellation of the decimal parts, that $N = 47/99$. This trick works for any two-digit pattern, that is, the nonterminating decimal $0.ababababababababababa\ldots$ equals $(10a+b)/99$. We need $10a+b$ in the numerator since the digit a is shifted into the tens place when we multiply by 100. This trick can be extended to any *repeating* nonterminating decimal. Thus we see that such decimals are *rational numbers*. Amazing! It is much harder to prove that nonterminating decimals that are not "repeating" are *irrational*.[4] The decimal expansion for $\sqrt{2}$, for example, has

---

[4]In the TV series *Star Trek*, Spock defeats an evil computer by asking it to compute the last digit of π. Since π is irrational, it does not have a "last" digit.

no repeating pattern. Neither does π, no matter what you've seen in the movies.[5]

There are many mysteries concerning infinite series that still puzzle mathematicians today. If you are under the mistaken impression that mathematics is dull and dry, perhaps you should read more about this intriguing subject and discover its hidden beauty.

The promise of the High Middle Ages – the energy and vitality, the heightened interest in learning, building, trading, and creating art – was, unfortunately not realized in the fourteenth century (for the most part) due to two tragedies. The Hundred Years War (1338-1453) sapped the strength of Europe and the dreaded Black Plague wiped out at least a quarter of her population. This horrible disease, believed to have originated in China, arrived in Italy in 1347 and spread like wildfire throughout Europe. (An ironic side effect was that labor became scarce and workers who survived the plague commanded higher wages.)

In spite of these setbacks, we shall soon see that an unstoppable outburst of empiricism and discovery was in the making that, by the fifteenth century, would lighten the way into the modern age of science and technology.

---

[5] π, the movie!

EXERCISES

1. Fibonacci first published his now-famous sequence as a problem. Suppose a one-month-old pair of rabbits (male and female) are too young to reproduce but are able to reproduce at the end of two months. Also assume that every month, starting from the second, they produce a new pair of rabbits (male and female). If each pair of rabbits reproduces in an identical manner, show that the number pairs of rabbits at the beginning of each month is a Fibonacci number.

2. Consider the following equations: $1 = 2-1$, $1 = 3-2$, $2 = 5-3$, $3 = 8 - 5$, $5 = 13 - 8$, and $8 = 21 - 13$. Adding the left and right sides of all of these equations yields the equation $1 + 1 + 2 + 3 + 5 + 8 = 21 - 1$. Show that this is always the case. In another words, show that the sum of the first n Fibonacci numbers is the $(n + 2)$nd term minus 1. In symbols, $u_1 + u_2 + u_3 + \cdots + u_n = u_{n+2} - 1$.

3. An *arithmetic sequence* is one for which there is a constant difference between consecutive terms. For example, if we start with the first term $u_1 = 2$ and a difference $d = 5$ we get the sequence $2, 7, 12, 17, 22, 27, \ldots$. Determine the first six terms for each of the following:

   (a) $u_1 = 3$, $d = 4$ (b) $u_1 = 1$, $d = 3$ (c) $u_1 = 0$, $d = 2$
   (d) $u_1 = 5$, $d = 5$ (e) $u_1 = 7$, $d = 7$ (f) $u_1 = 3$, $d = 5$

4. Show that the general nth term of an arithmetic sequence is given by the formula $u_n = u_1 + (n-1) \times d$.

5. A *geometric sequence* is a sequence in which each term is obtained from the previous one by multiplying by a fixed nonzero constant r, called the *ratio*, since it is the ratio of consecutive terms. For example, if we start with $u_1 = 1$ and $r = 2$, we get $1, 2, 4, 8, 16, 32, \ldots$, or, more simply, the powers of two. Determine the first six terms for each of the following:

   (a) $u_1 = 1$, $r = 3$ (b) $u_1 = 2$, $r = 3$ (c) $u_1 = 3$, $r = 2$
   (d) $u_1 = 1$, $r = \frac{1}{2}$ (e) $u_1 = 1$, $r = -2$ (f) $u_1 = 4$, $r = -\frac{1}{2}$

6. Show that the general nth term of a geometric sequence is given by the formula $u_n = u_1 \times r^{n-1}$.

7. List all lines of five boys and girls so that no two boys are consecutive. Verify that the answer is a Fibonacci number.

8. **A Fibonacci Trick.** Start with any two numbers and generate a sequence using Fibonacci's recurrence relation $u_{n+2} = u_n + u_{n+1}$. Confirm that the sum of the first ten terms is equal to 11 times the seventh term; in symbols this would be $u_1 + u_2 + u_3 + \cdots + u_{10} = 11 \times u_7$. For example, if we start with $u_1 = 3$ and $u_2 = 7$ we get the sequence 3, 7, 10, 17, 27, 44, 71, 115, 186, 301, and so on, and the sum of the first ten terms is 781 which is exactly 11×71. Prove that this happens for any starting numbers a and b.

9. Use the quadratic formula to solve the following quadratic equations:

   (a) $x^2 + 5x - 6 = 0$ (b) $x^2 + 4x + 4 = 0$
   (c) $x^2 + x - 6 = 0$ (d) $2x^2 + 3x - 10 = 0$
   (e) $x^2 - 100 = 0$ (f) $5x^2 + 10x + 4 = 0$
   (g) $x^2 = 3x + 10$ (h) $x(x-1) = 30$

10. Find the sum of the infinite series $1 + \frac{1}{3} + \frac{1}{9} + \frac{1}{27} + \frac{1}{81} + \cdots$.
    To do this, first denote the sum of the number

    $$1 + \frac{1}{3} + \frac{1}{9} + \frac{1}{27} + \frac{1}{81} + \cdots + \frac{1}{3^n}$$

    by S_n. (The denominators are all powers of 3.) Then, compute $3 \times S_n$ and subtract the equation for S_n to get an equation for $2 \times S_n$. Next, take the limit as n goes to infinity and, finally, divide by 2 to get the sum of the infinite series. Good luck.

11. Convert each of the following repeating decimals into a rational number. Be sure to reduce your answers.

    (a) 0.888888888888...   (b) 0.232323232323...
    (c) 0.450450450450...   (d) 0.901890189018...
    (e) 0.545454545454...   (f) 2.121212121212...

12. It is interesting to know that because π is irrational it must be approximated when trying to use it, say, to build something. Many of you may have used the rational number $\frac{22}{7}$ for π, but this is only an approximation. The ancient Egyptians used $\frac{256}{81}$, by the way. Using a calculator, determine the decimals for both fractions and determine which is the better approximation for $\pi = 3.1415926535\ldots$.

13. Throughout history many people have tried to compute the digits of π. The Old Testament tells us that π is 3. The Babylonians used $\frac{25}{8}$ for π. The Egyptians used $\frac{256}{81}$. Archimedes used $\frac{211875}{67441}$. Ptolemy used $\frac{377}{120}$ and Brahmagupta used $\sqrt{10}$. Compare these approximations for π with the infamous $\frac{22}{7}$. Also, compare these with the approximation $\frac{355}{113}$ by the Chinese mathematician and astronomer Tsu Ch'ung Chi. Who had the best approximation for π?

Suggestions for Further Reading

1. Beckman, Petr. *History of* π. St. Martin's Press, New York, 1976.

2. Gardiner, A. *Mathematical Puzzling*. Oxford University Press, New York, 1989.

3. Hoffman, Paul. *Archimedes' Revenge*. Ballantine Books, New York, 1988.

4. Muir, Jane. *Of Men and Numbers: The Story of Great Mathematics*. Dover, New York, 1996.

5. Pappas, Theoni. *The Joy of Mathematics*. Wide World Publishing/Tetra, San Carlos, CA, 1989.

6. Wells, David G. *You are a Mathematician: A Wise and Witty Introduction to the Joy of Numbers*. John Wiley & Sons, New York, 1997.

Chapter 7

Mathematics Marches On

In the 1100s in Paris, an innovative experiment was brewing.[1]
After five hundred years of Gregorian chants which featured a group
of people singing a single melody, attempts were made to make music

[1]No, not beer – the monks preferred champagne.

more interesting by dividing the singers into two groups and assigning to each a different melody!

The idea was simple and brilliant – the hard part was deciding what notes to give the second group. The first group, of course, sang the original melody. To give you an insight into the way this was, and is, done, we must speak briefly about music theory.

Now the entire subject of music is very mathematical, as you shall see. The basic notes of Western music consists of the C scale, whose notes c, d, e, f, g, a, b, c ascend in pitch. No doubt, you have sung this at some point and you might even have used the names do, re, mi, fa, sol, la, ti, do, as immortalized in the movie *The Sound of Music*, in the amusing song which describes "do" as a female deer.[2] These notes are, by the way, the white keys on a piano keyboard.

Notes are written today as circles on a staff of five lines. The location of the circle indicates the pitch – the higher the note, the higher the pitch. This is a profoundly mathematical idea! The range from the initial c to the final c is called an *octave*, the *oct* for the eight notes between the two c's (inclusive). The octave is also a measure of distance between the two notes. If you sing the notes, one after another, they will sound very similar. Other pitches are supplied by raising or lowering these pitches by a small amount, denoted by sharps (#) and flats (♭).

After the scale is finished, it starts again, until we reach the c two octaves above the one with which we started. This process occurs in both directions (up and down). The reference point is middle c, easily located in the middle of the piano keyboard. A piano has 88 keys – an impressive range indeed. The notes on the extreme left side sound very deep, while the notes on the extreme right sound shrill. This system did not evolve overnight. It took centuries. Notation in the Middle Ages was very primitive and spanned little more than the octave.

Next problem – the duration of the note, that is, the length of time it should be sustained by the singer or instrumentalist, is indicated by the way the circle is drawn. The unit of length is the so-called *quarter note*, denoted (no pun intended) as a filled-in circle with a stem. The composer can state the time value of a quarter note with reference to a metronome – a device that makes noise at constant time intervals. A quarter note can be shortened to half its value

---

[2]Many musicians use the solfege note si instead of ti.

by adding a small flag to its stem, thereby changing it to an *eighth note*. Note that 1/8 is half of 1/4. The next subdivision, which you may have anticipated, is the *sixteenth note*, written with two flags attached to its stem. One can continue this bisection process by adding more flags, each flag halving the previous time value.

In the opposite direction, a note whose duration is twice the value of a quarter note requires leaving the circle unfilled. If, in addition, the stem is removed, the result is a *whole note* – whose duration is four quarter notes. (Four quarters equals one whole, logical?) How does one obtain a 3/4 note, that is, a note of duration the same as three quarter notes? One widely used method places a dot after the unfilled circle of a half-note. This tells us to augment the value of the note by 50%, or, in other words, to add on half its value.

Another major issue is *meter*. Do we wish to group the beats in two's, three's, four's, and so on? Most rock songs use four quarter notes per measure. We call this signature "four-four" and display a brief sample in Figure 7-1. Measures are separated by thin vertical bars. Notice the "four-four" time signature at the beginning of the melody. A waltz usually has a "three-four" time signature, signifying that each measure contains three quarter notes.

What other qualities of music are measurable, and hence, are mathematical characteristics? The volume of a piece of music some-

Figure 7-1

times varies. This is measured with letters, such as *p* for *pianissimo*, meaning very quietly, and *f* for *fortissimo*, meaning very loudly. This is not adequate for indicating a gradual change in volume. This is achieved using pairs of diverging lines for an increase and converging lines for a decrease in volume, as illustrated in Figure 7-2.

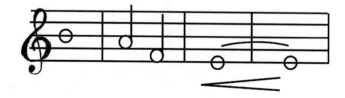

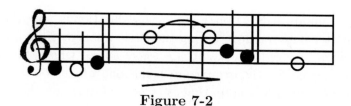

Figure 7-2

Finally, how do we measure the distance between two notes? How far away are c and g? Since we must count five notes to get from c to g, we call the interval between them a fifth. The notes c and f determine a *fourth*, since we must count four notes to get from c to f, and so on. The octave, fifth, and fourth are among the early intervals used in the twelfth-century attempts in creating a second melody to be sung simultaneously with the first. These early attempts were quite crude by later standards and over the next two hundred years were greatly improved by the inclusion of thirds and sixths. In the fourteenth century, composers started writing a bit of secular music, some for wealthy patrons, though the majority of works were religious in nature. In addition to the inclusion of pleasing intervals such as

thirds, composers started writing pieces with three or more simultaneous voices. The emphasis was often on creating independent voices
– accomplished by syncopating rhythms and varying the direction of
the melody, that is, one voice ascending and the other descending or
one voice sustaining a note while the other melody contains several
shorter notes. This kind of writing is called *counterpoint* and is very
difficult.

One of the results of the newly emerging music, with its pretty
harmonies, was a heightened awareness of the beauty of this world
and an increased desire to study it. In a sense, this typified the
spirit of the Renaissance. It should be noted that the Ancient Greeks
developed music theory to some degree, and it is from them that we
inherited our scales and letter notation. The Greeks used letters
like α and β to represent notes over a thousand years before the
Parisian musicians of 1100 did. They used dots over letters to denote
short duration, and dashes above them to indicate long duration – a
practice still adhered to today.

The art of the Middle Ages was as primitive as the music of the
times. As the drawing in Figure 7-3 illustrates, the canvas did not
capture what the eye sees. The circle over the angel's head should
be an ellipse, and the man in the background should be smaller than
the one in the foreground.

Figure 7-3

This typifies the way paintings appeared before the introduction of *perspective* in the Renaissance. With the influx of classical Greek geometry, it was not long before artists such as Fil-
ippo Brunelleschi (1377-1446), Leone Battista Alberti (1404-1472),
Piro della Francesca (1420-1492?), Albrecht Dürer (1471-1528), and
Leonardo da Vinci (1452-1519) began employing techniques such as
a vanishing point (or points) on the horizon line, to which receding
lines appear to converge.

No doubt, you have observed the seeming convergence of railroad
tracks as they recede into the distance, as if they meet on the horizon.
On the other hand, the railroad ties remain parallel – though they
appear to get closer to one another as they get further away from
the observer. Vertical telephone poles along the route remain parallel
but seem to get shorter.

Figure 7-4 shows a *two-point perspective* drawing of a cubic structure (a box in plain English) and includes the horizon line and the
two vanishing points on it.

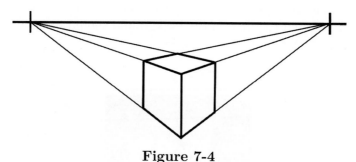

Figure 7-4

Clever methods were often employed to determine where to place
various lines in a drawing. Figure 7-5(a) shows a plaza with a four-by-
four array of large square tiles. In the one-point perspective drawing
of Figure 7-5(b), the five receding lines meet in the vanishing point,
P, on the horizon line, L. The line AB in the original figure is drawn
as the shorter line $A'B'$ in the perspective drawing. It is divided into
four equal segments by the five receding lines. The problem is, how
do we draw the three lines parallel to AB? More to the point (no
pun intended), how are the distances between them changing?

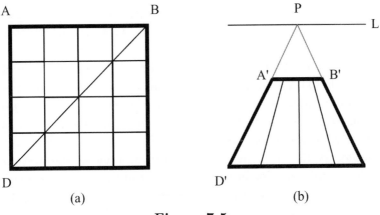

Figure 7-5

The ingenious solution is to draw diagonal DB and its image $D'B'$, as shown in Figure 7-6 (a) and (b). Since DB intersects the five receding lines where they meet the parallel lines, we can use the intersections in the perspective drawing to locate the parallels as is shown in Figure 7-6(b).

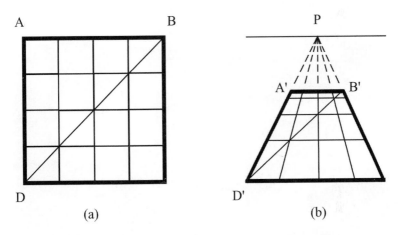

Figure 7-6

The Renaissance began in Florence, the cultural center of fourteenth-century Europe. Artists and intellectuals began to study the great literary, mathematical, and artistic achievements of the ancient glorious civilizations of Greece and Rome. Works were produced in the style of the classical masters. Dante (1265-1321)

wrote *The Divine Comedy* in a style reminiscent of Virgil. Bocac-
cio (1313-1375) and Petrarch (1304-1374) wrote prose and poetry
modeled after ancient works.

Much can be said about the spirit of the Renaissance, typified
by Leonardo Da Vinci. Though his friends did not refer to him as a
"Renaissance man," he certainly was the prototype. He was an artist,
scientist, inventor, architect, and mathematician, among many other
things. His curiosity encompassed all of nature and then some. He
saw geometry in the things he drew and was one of the first to see
the role of mathematics in the scientific study of the universe.

In the middle of the fifteenth century, the development of Johan
Gutenberg's "movable block" printing press ushered in a new era in
the transmission of knowledge as dramatic as the computer revolution some five hundred years later. By the end of the century, mass
production of texts on a host of subjects enhanced the distribution
of knowledge to the far reaches of the continent. In 1494, one of the
first printed mathematics text *Summa*, by Luca Pacioli (1445-1517),
contained a summary, as the title suggests, of virtually all of the
mathematics of the times.

In 1453, the Eastern Roman Empire fell to the Seljuk Turks,
which resulted in a mass exodus of Byzantine intellectuals to the
Italian city-states where the Renaissance was in full swing. They
brought a slew of Greek manuscripts along with them, thereby intro-
ducing many classical Greek works in philosophy and mathematics

into the boiling cauldron of intellectual and artistic activity in Europe.

Finally, the discovery of the New World in the 1490s added to the exciting air of empiricism. There was an increased willingness to explore, take risks, and challenge authority – of the church or of ancient authorities such as Aristotle, often referred to as "the philosopher." After all, the New World wasn't mentioned in the Bible, so revelation was not the final word in the acquisition of knowledge. Of course, the Reformation wars between Protestants and Catholics in the following century did much to further compromise the authority of the Vatican and its ability to retard free inquiry. The successful challenge of papal infallibility encouraged empiricism and ultimately gave scientists a safer environment in which to conduct their business.

A practical outgrowth of the Age of Exploration was the need for mathematics to design better ships and to provide adequate navigation for ocean voyages.

The theoretical mathematics of the fifteenth and sixteenth centuries consisted mostly of improvements in algebra, such as the development of formulas that solve cubic equations. A cubic equation is similar to a quadratic equation but includes a term of the form ax^3. In other words the general cubic is $ax^3 + bx^2 + cx + d = 0$. The formula involves the coefficients of the terms and is quite lengthy.

A part of the weakness of the algebra of the ancient world was the absence of notation. There was no symbol for the basic operations of adding, subtracting, multiplying, dividing, or calculating powers and roots. They didn't even have an equals sign! An even more serious problem was the use of different symbols for the unknown quantity and its square. How is one supposed to factor the difference $x^2 - x$ into $x(x - 1)$ if we use one symbol for x (the unknown quantity) and another, say S, for its square? We can't see the factoring if we have $S - x$ instead of $x^2 - x$. This problem of notation took hundreds of years of gradual development until it was fairly good – just in time for the incredible seventeenth century – the so-called *heroic century* of mathematics.

1. Draw a building using (a) one-point perspective and (b) two-point perspective.

2. Since $x^n = xxxxx \cdots x$ which is the product of n x's, and $x^m = xxxx \cdots x$ (the product of m x's), explain why $x^n \times x^m = x^{n+m}$. Use this law to find the following:

   (a) $x^2 \times x^4$ (b) $x \times x^3$ (c) $x^5 \times x^3$
   (d) $3x^4 \times 6x^5$ (e) $10x^7 \left(4x^3 - 2x^6\right)$
   (f) $2x^6 \left(6x^{10} + 2x^4 - x\right)$ (g) $\left(4x^3 + x^9\right) x^8$

3. Explain why
$$\frac{x^n}{x^m} = x^{n-m}$$

   Use it to find

   (a) $\frac{x^3}{x^2}$ (b) $\frac{x^6}{x^2}$ (c) $\frac{8x^{10}}{4x^3}$
   (d) $\frac{6x^5}{2x^2}$ (e) $\frac{10x^5 + 6x^7}{2x^3}$

4. Evaluate $\dfrac{x^n}{x^n}$ using the subtraction of powers rule and then use common sense to find the value of x^0. The resulting nonzero answer baffles most math students. Some even think it means x degrees!

5. From the rule in Exercise 2 we have $x^0 \times x^n = x^n$. Explain how this shows that $x^0 = 1$.

6. Can you think of a reason why $x^{\frac{1}{2}}$ is given the value $\sqrt{x}$? (*Hint*: $\sqrt{x}$ means that number which when multiplied by itself yields x, i.e., $\sqrt{x}\sqrt{x} = x$. Can you explain the meaning of $x^{\frac{1}{3}}$?)

7. Consider the problem
$$\frac{x^3}{x^5}$$

Compare the answers you get using the subtraction of powers law and using old-fashioned canceling. Show that

$$x^{-n} = \frac{1}{x^n}$$

Be careful not to think that a negative power makes the entire expression negative.

8. Since
$$\left(x^m\right)^n$$

means multiply x^m by itself n times, explain why

$$\left(x^m\right)^n = x^{mn}$$

Use this law to find

(a) $\left(x^3\right)^5$ (b) $\left(x^2\right)^7$ (c) $\left(x^{10}\right)^4$ (d) $\left(x^5\right)^8$ (e) $\left(x^9\right)^7$

(f) $\left(x^{99}\right)^0$ (g) $\left(x^{\frac{1}{3}}\right)^6$ (h) $\left(x^{\frac{1}{3}}\right)^2$ (i) $\left(x^{-5}\right)^2$ (j) $\left(x^5\right)^{-2}$

9. Since $(xy)^3 = (xy)(xy)(xy) = xyxyxy = xxxyyy = x^3y^3$, show that, in general,

$$(xy)^n = x^n y^n$$

Now show that
$$\left(x^m y^n\right)^k = x^{mk} y^{nk}$$

How may you now prove the law that $\sqrt{xy} = \sqrt{x}\sqrt{y}$?

10. This chapter has tried to show the connection between mathematics and art and music. Other examples of this include the demonstration of nothingness or the empty set in such works as (a) John Cage's musical piece called *4'33"* in which the pianist sits silently at the piano without making a sound, and (b) the abstract painting by Ad Reinhardt appropriately entitled *Abstract Painting* which consists of a canvas painted entirely in black. Try to find other examples in art and music, and summarize your findings in a five-page paper.

11. The music of Frank Zappa is an example of music that uses nontraditional odd time signatures, like 13/8 or 11/8, as well as odd phrasing. Obtain some of his sheet music, if possible, or listen to some of his music. Some interesting compositions are "King Kong," "Peaches en Regalia," or "The Black Page." Write a short paper about his music. Include your opinions on what you find.

12. The art of M. C. Escher is considered by many to be very mathematical. Find some of his art work on the Internet and write a short paper including your evaluation of his work.

Suggestions for Further Reading

1. Garland, T. H., Kahn, C. K., Stenstedt, K. *Math and Music: Harmonious Connections*. Dale Seymour Publishing, Parsippany, NJ, 1995.

2. Hofstadter, Douglas. *Gödel, Escher, Bach: An Eternal Golden Braid*. Basic Books, New York, 1999.

3. Ivins, William M. *Art & Geometry*. Dover, New York, 1946.

4. James, Jamie. *The Music of the Spheres: Music, Science, and the Natural Order of the Universe*. Copernicus Books, New York, 1995.

5. Locker, J. L., *M. C. Escher*. Harry N. Abrams, Inc. New York, 1982.

6. Pedoe, Dan. *Geometry and the Visual Arts*. Dover, New York, 1976.

7. Rothstein, Edward. *Emblems of the Mind: The Inner Life of Music & Mathematics*. Avon Books, New York, 1996.

Chapter 8

A Few Good Men

A handful of brave men armed with the weapons of mathematics and courage toppled, in a span of a mere one hundred years, the entire geocentric model of the universe. The Polish astronomer Copernicus[1] (1473-1543) challenged the geocentric model of Ptolemy (the one with the epicycles) on the grounds that placing the sun at the center of the solar system and assuming that Earth revolves about the sun (and rotates around its axis) reduces the number of equations describing the motion of the planets from about eighty down to thirty.

His book *De revolutionibus orbium coelestium* appeared in 1543 after his death. The Vatican ignored the book as it only suggested that the mathematical model putting the sun at the center makes more sense. He didn't assert that this is the way things are.

At the time of publication of this first round in the cosmic battle, the major hero, Galileo, was not yet born. We shall get to him soon.

A Danish astronomer, Tycho Brahe[2] (1546-1601), patiently collected a mountain of astronomical data over a ten-year period. Upon his death, his assistant Johan Kepler[3] (1571-1630), whom he had taught to observe and then hypothesize, interpreted the data and formulated his three laws of planetary motion.

1. The planets revolve about the sun in elliptical orbits, with the sun at one focal point of the ellipse. (An ellipse has two focal points, or foci.)

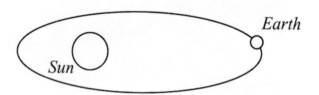

2. An imaginary line from the sun to a planet sweeps out equal areas in equal time intervals.

---

[1]Nicole Oresme also opposed the theory of a stationary Earth as proposed by Aristotle and advocated the motion of Earth some 200 years before Copernicus. He eventually rejected his own ideas.

[2]He was appointed Imperial Mathematician to the Holy Roman Emperor, Rudolph II, and Kepler was hired as his assistant to help with the calculations. He also wore a golden nose to replace his own which he lost in a duel.

[3]Kepler's mathematical work on the volume of a wine barrel is considered to be at the forefront of integral calculus and the calculation of volumes of solids of revolution.

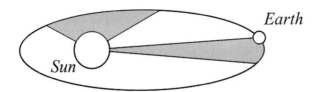

3. No matter which planet we study, the ratio of the square of the
   average distance from the sun to the cube of the length of time
   of one complete revolution is the same.

Aha! The motion of the planets is entirely predicable using mathematics. Furthermore, the Church and the ancient philosophers were wrong! The orbits are elliptic – not circular – and Earth is just another planet. And the best part is that these laws rest on mathematics and observation – not on authority. Kepler, too, escaped the wrath of the Roman inquisition. He lived outside of Rome's sphere of influence. Our main hero, as we shall see, was not as lucky.

Galileo[4] (1564-1642) was the son of a Florentine merchant. As a boy, he studied music, art, and poetry. He designed mechanical toys. He showed mathematical promise when he was a medical student.

---

[4]Immortalized forever in Queen's *Bohemian Rhapsody*.

He noticed that after the hanging lamps were filled with oil and lit, they swung back and forth in a periodic way. Although the arcs of these pendulums got progressively shorter, the amount of time for one full sweep back and forth remained constant. He filed this away in his young brilliant mind until, in his old age, he invented the grandfather clock based on the motion of a pendulum.

In 1589, he obtained a teaching post at the University of Pisa – a job that he soon lost because he spoke of errors he had found in Church cosmology. He sought employment in a college not under the sway of the Vatican.

Before long, he was a teacher in the secular University of Padua where he continued developing his interest in motion. The university was under the rule of Venice and, therefore, independent of papal authority.

Galileo was the first thinker to notice that the path of a thrown object is an upside-down parabola. This was interesting to the military folks who knew very little about the trajectories of cannonballs. Galileo showed them that the maximum range results when the angle of the cannon is 45°. Furthermore, a deviation from 45° in either direction results in the same shortening of the range. Thus an angle of 30° and an angle of 60° will produce the same range, though the trajectories will be different. (The common deviation here is 15°.) See Figure 8-1 to help you picture this.

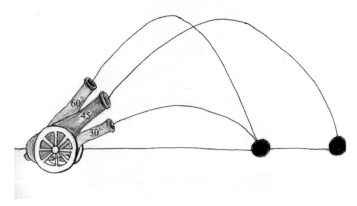

Figure 8-1

Galileo was a pioneer in the mathematical study of motion in other ways. He argued that since the distance traveled by an object moving with constant speed is the product of the speed and the time (*distance* = *rate* × *time*), one could compute distance by computing the area of a rectangle in a diagram, much like the graphs of today, in which a horizontal axis represents time and a vertical axis represents speed. A carriage traveling at 20 miles per hour for 10 hours clearly travels 200 miles, which is the area of the rectangle depicted in Figure 8-2, in which the axes are labeled t and v for time and velocity, respectively.

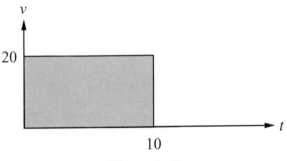

Figure 8-2

Now, said Galileo, suppose the carriage starts with velocity 20 miles per hour (in units of his day) and slows down at a constant rate to zero miles per hour at the end of the trip. In technical jargon, the carriage *decelerates uniformly*. Then we have the picture in Figure 8-3, and the resulting right triangle has area $\frac{1}{2} \times 10 \times 20 = 100$, meaning that the carriage travels 100 miles.

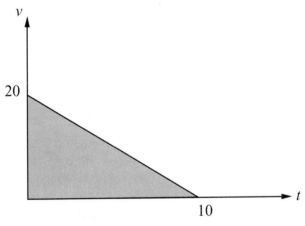

Figure 8-3

Uniform deceleration and acceleration interested Galileo because his experiments indicated that the velocity of a falling object accelerates at the rate of 32 feet per second, per second, that is, each second, the velocity increases by 32 feet per second. Physicists write this as ft/sec^2.

Galileo then proceeded to calculate the distance D that a dropped body falls in N seconds, as follows. (This argument is brilliant!) The chart of Figure 8-4 shows the speed of the falling object after N seconds.

time N	0	1	2	3	4
speed	0	32	64	96	128

Figure 8-4

Galileo then realized that the average speeds during these seconds are given by the table of Figure 8-5, where each average is obtained by adding the speeds at the endpoints of the time interval and dividing by 2.

time interval	1	2	3	4
average speed	16	48	80	112

Figure 8-5

Then the distance covered in N seconds is just the sum of the average speeds. Now Galileo saw the pattern of these average speeds. They are odd multiples of 16. That's right!

$$16 = 1 \times 16$$
$$48 = 3 \times 16$$
$$80 = 5 \times 16$$
$$112 = 7 \times 16$$

He then factored out the 16 and obtained, tentatively, that $D = 16\times$ (the sum of the first N odd numbers). Great! But what is the sum of the first N odd numbers? Fortunately, Galileo was familiar with the mathematics of ancient Greece. The Pythagoreans noticed that a square array of dots can be partitioned as a succession of L-shaped arrays of dots as shown in Figure 8-6. But each L has an odd number of dots and they are consecutive! Galileo's problem was solved two thousand years before he was born. The sum of the first N odd numbers is N^2. His formula $D = 16 \times N^2$ is still used in physics courses today – with different letters.

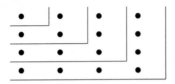

Figure 8-6

These types of geometric shapes and numbers fascinated the Pythagoreans. They called them *figurate numbers*. Another example of these, other than the *square numbers* given above, are the *triangular numbers* 1, 3, 6, 10, ... , which come from arranging dots in triangular arrays seen in Figure 8-7.

Figure 8-7

In 1609, Galileo heard about an invention from far away Holland that would dramatically affect the course of events in his life (and in ours) – the telescope. He immediately started to build one. One night in January 1610, Galileo turned his recently acquired telescope to the skies and observed several small white dots on one side of Jupiter (the planet – not the god). The next night, he saw two dots on that side, and the next night, saw one dot emerge on the other side of the planet. He immediately realized that he was seeing moons which were revolving around Jupiter – and not around the earth – in direct contradiction to the geocentric theory. He was also aware of another contradiction to Church (and ancient) cosmology. Aristotle and the church taught that heavenly bodies were perfect spheres. But "seeing is believing" – and Galileo saw craters on our very own moon.

He asked several scholars at the University of Padua to see for themselves. They refused. Their arguments were solid (to them).

(a) God would not create something that we couldn't see with the naked eye.

(b) There are seven heavenly bodies and no more. This is what the Bible tells us.

(c) All heavenly things rotate about the earth – not around other bodies.

He published his findings, of course. Then to make matters worse, in 1615, Galileo took his findings to Rome. We guess he didn't have good legal advice. He was told to go home and sternly warned to drop the matter. At this time, the Vatican decided to suspend Copernicus' book (then over sixty years old), realizing the magnitude of the challenge to their cosmology.

Galileo pressed his luck, as we say, and wrote a book entitled *Dialogues on the Two Principal Systems of the World* in which he ridiculed the Pope and the church-supported geocentric theory. He was eventually hauled before the Roman Inquisition, where he was given a chance to recant his heretical views and save himself from a horrible death at the stake. He recanted on his knees and spent the rest of his life between jail and house arrest. He wrote another book titled *Dialogues Concerning Two New Sciences*. This courageous man's writings were smuggled out of the country and published. He was pardoned by the Vatican posthumously in 1992. (This date is not a misprint.)

Galileo died in 1642, and the ailing geocentric theory died soon after. So did the philosophy of the medieval scholastics that motion must be explained using a *teleological* approach, that is, actions towards ends or goals such as "an apple falls to the ground because it belongs there," or "things are drawn to the earth because it is the most important place in the universe," or "the rose is red to symbolize the blood of Christ." Galileo established the validity of describing nature – mathematically of course. He substituted "how" for "why." This set a scientific trend for physics that is still valid today, although many theories include causal explanations. Galileo (and Kepler and others) showed that nature behaves in a mathematically predictable way, that is, he established the idea of natural law that we take for granted today but which was hardly obvious then.

Let us close this chapter with the observation of the English scientist Francis Bacon,[5] who dealt a final deathblow to medieval scholasticism which confined itself to *syllogistic*, or deductive logic, that is, proceeding from the general to the particular. He pointed out that science must be inductive. It must formulate general rules after studying particular facts.

Induction works mathematically. An alleged causal factor, A, must be present when the effect, B, is observed and must be absent when the effect is not observed. Moreover, A must be present to a greater degree when B is strong and ought to be present to a lesser degree when B is weak.

If one claims that it is the weight of a rock dropped on your toe that causes it to hurt, then a lighter rock should produce a quieter yelp of pain than a heavier one should. If one claims that it is the gray color that hurts, the claim is absurd since a gray pebble produces no pain while a heavy brown rock elicits a (barbaric) yelp.

The fifteenth and sixteenth centuries saw the world turned upside down – the Reformation, the new world, the printing press, Earth exiled to the third tier in the solar system. The sails of change were raised to the wind and the world was ripe for revolution!

---

[5]Knowledge is power – Nam et ipsa scientia potestas est – from *Meditationes Sacræ. De Hæresibus.*

1. Find the sum of the first 10 odd numbers in two ways:

    (a) by addition, and

    (b) by the ancient Greek method used by Galileo.

2. The first odd number is 1, the second is 3, the third is 5, and the fourth is 7. Find a formula for the nth odd number. (Look for a pattern with the first few odds.) What is the one-hundredth odd number?

3. Observe that $1 + 2 + 1 = 4$, $1 + 2 + 3 + 2 + 1 = 9$, and $1 + 2 + 3 + 4 + 3 + 2 + 1 = 16$. Notice that the sums are squares. Show that this pattern persists by dissecting a square array of dots in a clever way using diagonal cuts.

4. The Ancient Greeks were very interested in the triangular numbers 1, 3, 6, 10, 15, 21, 28, 36, 45, ... These numbers represent triangular arrays of dots. Bowlers recognize the ten pins formed by four rows of the consecutive numbers 1, 2, 3, and 4, which add up to 10. Pool players recognize the triangular number 15. The nth triangular number, denoted t_n, is $1 + 2 + 3 + \ldots + n$. Find the tenth, eleventh, and twelfth triangular numbers.

5. Show that the general formula for the triangular numbers is

$$t_n = \frac{n(n+1)}{2}$$

(*Hint*: Write the sum backward and forward and then deduce an expression for $2 \times t_n$. Then divide by 2.)

6. Observe that the sum of any two consecutive triangular numbers in the short list in Exercise 4 is a square, for example, $3 + 6 = 9$ and $15 + 21 = 36$.

   (a) Show that this is always true. (*Hint*: Cut a square array of dots into two triangular arrays of numbers, only one of which includes a full diagonal of dots.)

   (b) Prove this algebraically. (*Hint*: If the nth triangular number is

   $$\frac{n(n+1)}{2}$$

   write the preceding triangular number by substituting $n - 1$ for each n in this expression, yielding

   $$\frac{(n-1)n}{2}$$

   Now add these consecutive triangular numbers and show that the sum is indeed n^2.)

7. Observe that the squares of the first five triangular numbers are $1, 9, 36, 100$, and 225. Note that the successive differences are $9 - 1 = 8$, $36 - 9 = 27$, $100 - 36 = 64$, $225 - 100 = 125$. The numbers $8, 27, 64$, and 125 are the cubes of the integers $2, 3, 4$, and 5, respectively. This is not a coincidence. Can you prove that the square of the nth triangular number minus the square of its predecessor equals n^3?

8. Explain the average speeds Galileo computed for a falling body. Why are they odd multiples of 16?

9. The numbers $1, 5, 12, 22, \ldots$, are sometimes referred to as *pentagonal numbers* because these numbers represent pentagonal arrays of dots, as pictured below.

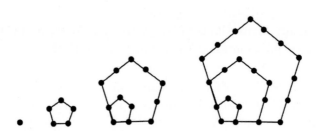

Continue the pattern and generate the first ten pentagonal numbers.

10. Show that the general formula for the nth pentagonal number is
$$\frac{n\,(3n-1)}{2}$$

Suggestions for Further Reading

1. Brody, D. E., and Brody, A. R. *The Science Class You Wish You Had: The Seven Greatest Scientific Discoveries in History and the People Who Made Them.* Berkeley Publishing Group, New York, 1997.

2. Copernicus, Nicolaus. *On the Revolutions of Heavenly Spheres,* Great Mind Series. Prometheus Books, New York, 1995.

3. Galilei, Galileo. *Dialogues Concerning Two New Sciences,* Great Mind Series. Prometheus Books, New York, 1995.

4. Guillen, Michael. *Five Equations that Changed the World.* MJF Books, New York, 2000.

5. Kepler, Johannes. *Epitome of Copernican Astronomy & Harmonies of the World,* Great Mind Series. Prometheus Books, New York, 1995.

6. Spielberg, N., and Anderson, B. *Seven Ideas That Shook the Universe.* John Wiley & Sons, New York, 1987.

Chapter 9

A Most Amazing Century of Mathematical Marvels!

Let's take a moment to get an overview of the role of mathematics at the beginning of the seventeenth century. Commerce, banking, insurance, astronomy, artillery, science, navigation, perspective in visual art, the calendar, mapmaking, music theory and instrument construction, architecture, and machine making are just a few things which come to mind! Nonetheless, we are about to study the century that surpassed all of the mathematics produced since the first time a caveman counted the number of members in a hunting party. Enter the cast of our play.

The first great hero of the seventeenth century is Pierre de Fermat[1] (1601-1665) – an amateur mathematician! He heard the seductive call of number theory and made many amazing discoveries in that time-honored field. He rarely proved his statements or at least rarely showed his proofs to others. Amazingly, though, his statements were hardly ever wrong. One of his assertions challenged mathematicians for almost four hundred years. The Princeton mathematician Andrew Wiles finally proved it in 1996. This assertion is commonly referred to as *Fermat's Last Theorem*, even though it wasn't a theorem until 1996. Fermat's Last Theorem states that the equation

$$a^n + b^n = c^n$$

has no solution for any integers a, b, and c, all different from zero, when n is greater than 2. If, for example, $n = 3$, this says that the sum of two cubes (like 8, 27, 64, 125, etc.) is *never* a cube.

---

[1]Fermat wrote his famous Last Theorem in the margins of *Arithmetica* by Diophantus and as the story goes the margin wasn't big enough for him to write his truly wonderful proof.

Another theorem of his (called *Fermat's Little Theorem*) says that if p is a prime number and a is a positive integer not divisible by p, then the number

$$a^{p-1} - 1$$

is a multiple of p. To illustrate this, let $p = 5$ and $a = 3$. Then $3^4 - 1$ is a multiple of 5. (Of course 80 *is* a multiple of 5.) The amazing thing is that this will always happen, as long as p is prime and a is not divisible by p.

Fermat sought a formula that would yield only the prime numbers. He thought it was given by raising 2 to the power 2^n and then adding 1. When $n = 1$, we get 5 which is prime. When $n = 2$, we get 17 – also a prime. The formula works when n is 3 or 4. So Fermat concluded that it always works. It took about one hundred years until the great Swiss mathematician Leonard Euler proved that he was wrong. When $n = 5$, we get a number divisible by 641. This was quite a feat considering that there were no calculators back then. Try to compute

$$2^{32} + 1$$

and then try to factor it without a calculator! Good luck.

Fermat founded a whole new subject – *modular arithmetic*. Though it took several centuries to develop the notation we shall now use, it was Fermat who started the ball rolling. This kind of arithmetic is very important in many fields today including computer science and cryptography.

Modular arithmetic starts with a so-called *modulus* n, which divides all numbers into n mutually exclusive *congruence classes* or *residue classes*. This number, n, must be set in advance. Let's use $n = 5$. This modulus divides all integers into 5 distinct residue classes modulo 5 by considering the remainder after division by 5. Clearly, the possible remainders are 0, 1, 2, 3, and 4. If the remainder is larger than 4, you divided incorrectly! All numbers that leave a remainder of 0 form a residue class whose so-called *least residue* is 0. The (positive) members of this class are 5, 10, 15, 20, and so on – in short, all multiples of 5, that is, all numbers of the form $5k$. On the other hand, all numbers that leave a remainder of 1 form another class (with least residue 1). This class contains 1, 6, 11, 16, 21, and so on – numbers expressible as $5k + 1$. This continues until the last class, with least residue 4, containing numbers of the form $5k + 4$ – such as 4, 9, 14, 19, 24, and so on.

Note that there is no class of the form $5k + 5$. Such a number can be factored into $5 \times (k + 1)$ which is clearly in the class $5k$, that is, it is $5 \times$ (something). By placing the integers in five columns, as shown below, we can visualize the five classes as columns in this matrix which only goes to 36. In reality, it continues forever in both directions (up and down).

	...	−3	−2	−1
0	1	2	3	4
5	6	7	8	9
10	11	12	13	14
15	16	17	18	19
20	21	22	23	24
25	26	27	28	29
30	31	32	33	34
35	36	...		

The least residues are in the first full row. The equation

$$a \equiv b \,(\mathrm{mod}\, 5)$$

indicates that the numbers a and b are in the same column, that is, in the same residue class modulo 5. As an example, we have the following.

Example 9-1 *Verify that* $16 \equiv 31 \, (\mathrm{mod} \, 5)$ *and* $18 \equiv 28 \, (\mathrm{mod} \, 5)$.

> Since $16 \div 5 = 3 \, R \, 1$ and $31 \div 5 = 6 \, R \, 1$ they belong to the same least residue class 1. Thus, $16 \equiv 31 \, (\mathrm{mod} \, 5)$. Similarly, $18 \div 5 = 3 \, R \, 3$ and $28 \div 5 = 5 \, R \, 3$. Therefore, $18 \equiv 28 \, (\mathrm{mod} \, 5)$. ∎

Do you see that whenever two numbers are in the same column, their difference is a multiple of 5? On the other hand, if two numbers are in different columns, their difference is not a multiple of 5. Mathematicians say these two sentences as follows. Two numbers are in the same column *if and only if* their difference is a multiple of 5.

Let's switch to a different modulus before you start believing that there is only one modulus.[2] As there are seven days in the week, let's consider this modulus. The residue classes, or remainders, when dividing by 7 are 0, 1, 2, 3, 4, 5, and 6. We would expect 7 classes since 7 is the modulus. The matrix below shows the residue classes. We start at 0 for convenience.

0	1	2	3	4	5	6
7	8	9	10	11	12	13
14	15	16	17	18	19	20
21	22	23	24	25	26	27
28	29	30	31	32	33	34
35	36	37	38	39	...	

Notice, again, that two numbers are in the same column if and only if their difference is a multiple of the modulus 7. We have, then, a general rule

$$a \equiv b \, (\mathrm{mod} \, n) \text{ if and only if } a - b \text{ is a multiple of } n$$

This works whether we use $a - b$ or $b - a$. They are negatives or opposites of one another. For example, if $a = 10$ and $b = 7$, then $a - b = 3$ while $b - a = -3$. They are both multiples of 3.

The modular equation $50 \equiv 1 (\mathrm{mod} \, 7)$ tells us that if today is, say Monday, then in 50 days, it will be Tuesday. In other words, the difference between 50 and 1, namely 49, is a multiple of 7 and

---

[2]Would such a person be called a monomodulist?

can, therefore, be ignored if we are interested only in the day of the week, and not in the number of days that transpire. Thus, modular arithmetic views the journey through the integers as a *periodic*, that is, a repeating one – almost like an ascent up a circular stairway. After ascending a certain number of steps, we are directly above where we started, and so on.

This is also similar to the progression of musical notes as we plunk the 88 keys of the piano from left to right. The seven notes "a, b, c, d, e, f, and g" keep on repeating. Thus an ascending melodic passage "e f g a b c" makes sense. The "a" after the "g" is an octave higher than the "a" a fifth below the beginning "e." Small wonder that mathematical and musical potential are as linked as they are in most people. Many professional physicists and mathematicians play musical instruments.

Back to modular arithmetic. Referring back to the matrix of numbers modulo 7, if we add any number in the 2 column to any number in the 3 column, notice that the sum is in the 5 column! For example, $9 + 10 = 19$ and $19 \equiv 5 (\bmod\, 7)$. What if we add numbers in the columns of 4 and 6, respectively? The answer will be in the 10 column – which is the same as the column headed by 3, since $10 \equiv 3 (\bmod\, 7)$. This same weird phenomenon occurs when we multiply. If $a \times b = c$, then the least residues of a and b modulo n will have a product equal to $c(\bmod\, n)$. In fact, modular arithmetic is almost exactly like ordinary arithmetic in the sense that the following laws apply to equations modulo n. Some of them will be obvious while some might require some thought to justify them. To save type, we shall omit the phrase $(\bmod\, n)$, but remember that it should appear in every equation below.

1. If $a \equiv b$, then $b \equiv a$. (*Symmetry law*)

2. If $a \equiv b$ and $b \equiv c$, then $a \equiv c$. (*Transitivity law*)

3. If $a \equiv b$ and $c \equiv d$, then $a + c \equiv b + d$ and $a - c \equiv b - d$.

4. If $a \equiv b$, then $ac \equiv bc$.

5. If $a \equiv b$ and $c \equiv d$, then $ac \equiv bd$.

6. If $a \equiv b$ and r is an integer, then $a^r \equiv b^r$

Be careful with cancellation! If $ac \equiv bc \pmod{n}$, it does not follow that $a \equiv b \pmod{n}$, that is, one can't cancel the c, except under certain circumstances. To illustrate this point, while it is true that $8 \equiv 12 \pmod 4$, it does not follow that one can cancel the 2, yielding the false equation $4 \equiv 6 \pmod 4$. The sharp reader[3] might notice that the statement is true if we cancel the 2 in the modulus as well, that is, it does follow that $4 \equiv 6 \pmod 2$.

In terms of modular arithmetic, Fermat's Little Theorem may be written

$$a^{p-1} \equiv 1 \pmod{p}$$

assuming, of course, that p is a prime number, and a is any number not divisible by p. Put another way, it is only necessary that a and p be *relatively prime*, that is, that they should have no common factor other than 1. Put even a third way, their greatest common divisor (GCD) must be 1.

Modular arithmetic goes hand in hand with another type of mathematical operation known as the *greatest integer function*. Modular arithmetic is interested in the remainder after dividing two numbers, while the greatest integer function is interested in the quotient (at least for positive numbers). For example, if we denote the greatest integer function of a number x by $[x]$, then the greatest integer of 22 divided by 4 is $\left[\frac{22}{4}\right]$ which is equal to 5 or, more simply,

$$\left[\frac{22}{4}\right] = 5$$

Modern computer programmers use modular arithmetic and the greatest integer function to convert units of one kind into a mixture of units of several kinds. Thus, to convert 80 inches into feet and inches, we first take the greatest integer function of 80 divide by 12 which is 6, that is, $\left[\frac{80}{12}\right] = 6$, giving us the number of feet. Then we observe that $80 = 8 \pmod{12}$, giving us the number of inches. Hence, 80 inches is 6 feet, 8 inches. Let's look at another example.

Example 9-2 *Convert 200 inches to yards, feet, and inches.*

We first take the greatest integer function of 200 divided by 36 to get the number of yards, $\left[\frac{200}{36}\right] = 5$ yards. Now,

---

[3] One with a slightly raised musical pitch.

the remaining number of inches is $200(\bmod 36) = 20$. For the number of feet, we use the greatest integer function of 20 divided by 12, which is $\left[\frac{20}{12}\right] = 1$ foot. Finally, the leftover $20(\bmod 12) = 8$ is the number of inches. Thus, 200 inches is 5 yards, 1 foot, and 8 inches. ∎

In modern programming languages, like C++ or Java, the command which yields the least residue of $m(\bmod n)$ is usually written $m \% n$. This is a useful way of telling the computer how many numbers to put into a row of a matrix, or array, before going to the next row. If you wish to enter the first 100 numbers in ten rows, you would need to skip to the next line after 10, 20, 30, and so on. If n counts the numbers, the command in C reads

```
if (n%10==0) printf(''\n'');
```

This means, if $n \equiv 0(\bmod 10)$, that is, if n is a multiple of 10, execute the "new line" command \n, as it is called. We will discuss the miracle of computers in the final chapter.

Our next seventeenth century superhero is Blaise Pascal[4] (1623-1662) – one of the founders of probability theory.[5] Let us begin at the beginning. Suppose we must perform two tasks in succession. Assume, moreover, that the first task involves choosing one of m options, and the second involves choosing one of n options. Then there are $m \times n$ different ways to perform the two tasks. As an example, consider the following.

Example 9-3 *Suppose you are ordering dinner at the Binomial Cafe (so called because you should "buy no meal" there.) The menu is given below In how many ways can you order a full dinner?*

---

[4]Pascal once wrote, "Men never do evil so completely and cheerfully as when they do it from religious conviction."

[5]Pascal corresponded with Fermat about many ideas which led to the theory of probability.

The Binomial Café Menu

Overpriced Appetizers

1. Cheese with Crackers
2. Cheese without Crackers
3. Crackers without Cheese

Small Main Dishes

a. Morsel of Meat
b. Sautéed mushroom au gratin

A service charge of 25% will be added to the already outrageous subtotal.

There are 3 ways to order an appetizer and there are 2 ways to order a main dish. Assuming a full dinner consists of an appetizer and a main dish, there are $3 \times 2 = 6$ ways to order dinner. This also implies that if someone tries to guess what you had for dinner, their chances are only 1 out of 6. ∎

As another illustration, say you are traveling from New York City to San Francisco, with a stopover in Chicago. The first part of your great trek across America can be done either by train, bus, rented car or plane. The second can be done only by train or bus, because you are afraid to fly over the Rocky Mountains and can't afford an extended car rental. Then there are 4 ways to complete the first leg of your journey and 2 ways to triumphantly complete the second. By the $m \times n$ law, there are 8 possible itineraries.

In these cases, one can actually list all the possibilities to verify the answers. Letting T, B, C, and P stand for train, bus, car, and plane, we obtain the following possibilities:

TT	TB
BT	BB
CT	CB
PT	PB

The four rows correspond to the four possibilities (T, B, C, and P) for the first leg of the journey, while the two columns correspond to the two methods (T and B) for completing the second leg. This is why the law tells us to multiply! The number of cells in a box with m rows and n columns is precisely $m \times n$.

What if the journey has three legs? The law extends in just the way you would expect. If the number of ways of performing three tasks are m, n, and r, then there are $m \times n \times r$ ways to perform all three tasks. Why stop here? In how many ways can five people line up in front of a bank teller? The first place can be filled in five ways. This leaves only four people to fill the second place slot. The third place can be filled in only three ways, while the fourth place has only two possibilities. The last place can be filled in only one way. Using an extended $m \times n \times \cdots$ law, called the *multiplication rule*, we get $5 \times 4 \times 3 \times 2 \times 1 = 120$ ways!!!!!! "But there are only five people!" you say. We invite you to use the letters a, b, c, d, and e to represent the five people on line at the bank, and we promise you 120 different configurations. We'll start you off.

abcde, abced, abdce, abdec, abecd, abedc, ...

You may recall the definition of n factorial, written $n!$ in mathematical language. This should not be read with emphasis. It simply stands for $n \times (n-1) \times (n-2) \times \cdots \times 2 \times 1$.

The first few factorials are listed in the table below.

$$1! = 1$$
$$2! = 2$$
$$3! = 6$$
$$4! = 24$$
$$5! = 120$$
$$6! = 720$$
$$7! = 5040$$
$$8! = 40,320$$
$$9! = 362,880$$
$$10! = 3,628,800$$

We can now state the following law, in which the term *permutation* means *linear arrangement* or *order*.

Law: There are $n!$ permutations of n distinct objects.

It never ceases to amaze us that there are over 3,000,000 ways for ten people to form a line. Put another way, there are over 3,000,000 ways to permute the letters *abcdefghij*. Factorials grow very quickly. There are only six ways to permute the letters *abc*. We can even verify this by listing them: *abc, acb, bac, bca, cab, cba*. We don't recommend this for permutations of *abcdefghij*, but you're free to try.

New (but related) problem. Given n distinct objects, in how many ways[6] can we select k of them, ignoring the order in which the k objects are selected? Assume, for example, that you have five friends, so $n = 5$. You have two extra tickets for the Broadway musical "Oh, Calculator"[7] and would like to choose two of your friends to accompany you. In how many ways can this be done? Let's call your friends a, b, c, d, and e. Let's list all the possibilities:

ab	ba	ca	da	ea
ac	bc	cb	db	eb
ad	bd	cd	dc	ec
ae	be	ce	de	ed

---

[6] Paul Simon conjectured, "there must be fifty ways to leave your lover" but listed only five.

[7] A touching story about a boy and his calculator.

We could have predicted this using the multiplication rule. There are 5 ways to pick the first friend and only four ways to pick the second, so the principle yields $5 \times 4 = 20$, which is in perfect agreement with the twenty entries in the matrix above.

But ... there is a huge mistake in the matrix!! Each possibility is counted twice! The first entry, ab, (say Albert and Bartholomew) is the same as the next entry in the first row – ba (Bartholomew and Albert). We are not concerned with the order (permutation) of the two people selected. We just want to know who the lucky ones are. We must divide our answer, 20, by 2, yielding the correct answer – 10. We call this "the number of combinations of five objects taken two at a time," or more simply "five choose two," and write this as $C(5,2)$, $_5C_2$, or $\binom{5}{2}$. We will use the notation $C(5,2)$. More generally, $C(n,k)$, represents the number of combinations of n distinct objects, taken k at a time.

Let's figure out a formula for this strange variable number $C(n,k)$, or "n choose k." This means that we are given n distinct objects and we wish to know how many ways there are of selecting k of them. Suppose, for example, we have ten different novels at home and wish to take three of them along on a vacation. This can be done in $C(10,3)$ ways. Great. But what is $C(10,3)$? We can choose the first novel in 10 ways, the second novel in 9, and the third in 8, implying that there are $10 \times 9 \times 8$ (wow – 720) ways to list three out of the ten novels. But ... each combination is listed 3!, or six, times. This means that we must divide 720 by 6, yielding 120. We can now obtain a general procedure for computing $C(n,k)$. Begin by computing the first k factors of $n!$. Then divide by $k!$. This is exactly what we just did for $C(10,3)$. We computed

$$\frac{10 \times 9 \times 8}{3!} = \frac{720}{6} = 120$$

How do we write this out mathematically? Here it comes:

Law: The number of combinations of n distinct objects taken k at a time, denoted $C(n,k)$, equals

$$\frac{n \times (n-1) \times (n-2) \times \cdots \times (n-k+1)}{k!}$$

It is sometimes convenient for theoretical purposes to multiply the top and bottom of this last fraction by $(n - k)!$ which turns the numerator into $n!$ by supplying the remaining factors. We then get the alternative formulation

$$C\left(n, k\right) = \frac{n!}{k!\left(n - k\right)!}$$

But what if order of selection matters? If we wish to choose a president and vice-president for the "Society to Abolish College Mathematics Requirements" and we have five eligible candidates, say a, b, c, d, and e, we must now take order into account! The first letter shall represent the president and the second, the vice-president. Thus, ab and ba have different meanings, and both must be counted, bringing the total (using the multiplication rule) to 5×4, or 20. The law here simply dispenses with the denominator $k!$ in the above formula for $C(n, k)$. We now have the following law regarding permutations.

Law: The number of permutations of n distinct objects taken k at a time, denoted $P(n, k)$, equals

$$n \times (n - 1) \times (n - 2) \times \cdots \times (n - k + 1)$$

Returning to the numbers $C(n, k)$, we have an interesting observation. If we choose k people from a pool of n eligibles, isn't this the same as choosing to pass on $n - k$ people? Inviting two out of five friends means choosing the three friends who will be left out. Cutting to the chase, $C(n, k) = C(n, n - k)$. Thus, the nightmare calculations required to compute $C(100, 97)$ need not be done, since $C(100, 97) = C(100, 3)$ and this last expression is simply

$$\frac{100 \times 99 \times 98}{3!} = \frac{970,200}{6} = 161,700$$

What is the point of all this? Mathematicians of the seventeenth century were interested in a new kind of mathematics, called *probability theory*.[8] The probability of an event is defined to be the number

---

[8]Levi ben Gerson (1288-1344) actually developed some of the ideas used in probability theory hundreds of years before Pascal.

of successes divided by the number of all possible outcomes, assuming that all outcomes are equally likely.

If we toss a fair coin, there are two (equally likely) possible outcomes, heads or tails. Then the probability of getting heads is one divided by two, or $\frac{1}{2}$. This does not mean that if you toss a coin ten times, you will get exactly five heads each time. It does mean, however, that if you toss a coin n times, the number of heads will approach the expected number, that is, $\frac{n}{2}$, as n goes to infinity.

One more example. If we toss a pair of dice and wish to get 7 or 11, what is the probability of success? The chart below shows us the 36 possible (equally likely) outcomes.

+	1	2	3	4	5	6
1	2	3	4	5	6	**7**
2	3	4	5	6	**7**	8
3	4	5	6	**7**	8	9
4	5	6	**7**	8	9	10
5	6	**7**	8	9	10	**11**
6	**7**	8	9	10	**11**	12

The eight winning numbers 7 and 11 in bold indicate that the probability of a win is $\frac{8}{36}$, which can be reduced to $\frac{2}{9}$ or, in decimal form, $0.222\ldots$. This can also be expressed as a percent by shifting the decimal point two places to the right, yielding approximately 22.2%.

A decimal can always be converted to a percent using this trick. The fraction $\frac{1}{3}$, for example, is $0.333\ldots$ in decimal form and is approximately 33.3%. A fraction may be converted directly into a percentage by multiplying it by 100. Thus $\frac{1}{2}$ becomes $\left(\frac{1}{2} \times 100\right)\%$, or 50%, while $\frac{1}{4}$ becomes $\left(\frac{1}{4} \times 100\right)\%$ or 25%. On the other hand, a percentage may be converted into a fraction by dividing by 100. Thus 44% becomes $\frac{44}{100} = \frac{11}{25}$. This is what *percent* means – out of a hundred. One can convert a percentage into a decimal and a decimal into a percent by shifting the decimal point two places to the left or right, respectively, adding on zeros if necessary. So 8.25%, which can be written 008.25%, is equivalent to the decimal number 0.0825 and the decimal 1.7, which can written 1.700, is 170.0% or just 170%.

The probabilities $\frac{1}{2}$ and $\frac{2}{9}$, obtained above for getting heads and 7 or 11, respectively, are examples of theoretical probabilities. They are obtained without actually having to flip a coin or toss a pair of dice. They require pencil, paper, and brain. The Ancient Greek mathematicians would have approved! What a triumph of pure thought!

Many probabilities are derived empirically, that is, as a result of observation and counting. If a college has a population of $5,000$ students and if 500 of them are mathematics majors, we obtain the empirical probability $\frac{500}{5000}$, or more simply, $\frac{1}{10}$, of a randomly chosen student being a math major. To get this probability, we must do some legwork and gather statistics. Now here is an important question. If 20 homes in a community of 200,000 homes burned down *last* year, can we assume that the probability of a home burning down *this* year is $\frac{20}{200,000}$? If the numbers are obtained empirically, will they continue to be valid? This is incredibly important to insurance companies as it helps them determine premiums. This idea grew in the seventeenth century when merchants in the Dutch East India Company wanted to insure their freight against losses at sea, which were quite substantial. The mathematics is ingenious! Everyone pays a premium determined by an insurance company by keeping careful track of yearly losses and noticing their approximate regularity. Thus each merchant pays – even if he has no losses that year – while no merchant is wiped out by tragedy at sea.

Let us put this aside to dry (no reference to losses at sea here) for awhile. We shall return to it in a very dramatic way. Recall that $(a+b)^2 = a^2 + 2ab + b^2$. If we multiply this answer by $a + b$ and collect *like* terms, we get $(a+b)^3 = a^3 + 3a^2b + 3ab^2 + b^3$. If you have the patience to multiply one more time by $a + b$ and simplify, you will get $(a+b)^4 = a^4 + 4a^3b + 6a^2b^2 + 4ab^3 + b^4$. Notice that the first and last coefficients in each expression for $(a+b)^n$ are 1, which we usually omit. After all, it seems like a waste of time to write $1x$ for x. The 1 is understood. Before we get to the point, $(a+b)^0 = 1$ and $(a+b)^1 = a + b$, right? Now we will get to the point. Mathematicians were aware of the following triangle, now known as *Pascal's triangle*,[9] obtained by listing the coefficients of the expansions of powers of $a + b$:

$$
\begin{array}{ccccccccccccc}
 & & & & & & 1 & & & & & & \\
 & & & & & 1 & & 1 & & & & & \\
 & & & & 1 & & 2 & & 1 & & & & \\
 & & & 1 & & 3 & & 3 & & 1 & & & \\
 & & 1 & & 4 & & 6 & & 4 & & 1 & & \\
 & 1 & & 5 & & 10 & & 10 & & 5 & & 1 & \\
1 & & 6 & & 15 & & 20 & & 15 & & 6 & & 1 \\
\end{array}
$$

After examining the entries, you might notice that each entry is the sum of the two entries immediately above it (to the left and right). The 15 in the last row of the chart, for example, is the sum of the 5 and 10 in the row above. Each of the infinitely many rows of Pascal's triangle is named after its second entry. Thus the last row in the above chart is the 6th row, and not the 7th as a strict count would show.

The amazing fact is that the entries of the nth row are the numbers

$$C(n,0), C(n,1), C(n,2), \ldots, C(n,n-2), C(n,n-1), \text{ and } C(n,n)$$

The numbers in the fourth row, 1, 4, 6, 4, and 1, for example, are $C(4,0), C(4,1), C(4,2), C(4,3),$ and $C(4,4)$!! This is as good a time

---

[9]Evidence shows this to have been known well before Pascal, but he is the one lucky enough to have his name attached to it.

as any to explain $C(n,0)$ and $C(n,n)$ (which are equal, by the way, since $0 + n = n$). Recall the formula given above for $C(n,k)$ involves division by $k!$. What could we possibly mean by *zero factorial*?

Mathematicians were forced to define this as 1 for consistency. Then, using the alternative formulation

$$C\left(n,k\right) = \frac{n!}{k!\left(n-k\right)!}$$

we get

$$C\left(n,0\right) = \frac{n!}{0!\left(n-0\right)!} = \frac{n!}{n!} = 1$$

This is actually very logical. There is only one way to select no one from a group of n people – simply walk away! On the other hand, the number

$$C\left(n,n\right) = \frac{n!}{n!\left(n-n\right)!} = \frac{n!}{n!0!} = \frac{n!}{n!} = 1$$

is again quite logical. There is only one way to select n people from a pool of n candidates – choose them all.

If you like patterns, Pascal's triangle is full of them. The sums of the elements of each row in the chart above are 1, 2, 4, 8, 16, 32, and 64 – the powers of two! In fact, the sum of the entries of the nth row is precisely 2^n. A quick way to establish this is to observe, firstly, that it works for the first few rows, and then, secondly, that each entry in a given row is counted twice in the next row because of the pattern of generation of the triangle. It appears in the entry immediately to its left (in the next row) and immediately to the right. Then row sums must double. Incredible!

Readers still hungry for patterns might notice that the third entry in each row yields the triangular numbers (1, 3, 6, 10, and 15 in the chart above). Is this just a coincidence? Mathematicians rarely believe that a pattern is coincidental.

Our next great hero is the philosopher René Descartes (1596-1650). In 1637 he published a philosophical work with an appendix dealing with coordinate geometry.[10] This was one of the most important achievements of the seventeenth century. His book had one of the longest titles of the seventeenth century: *Discours de la méthode pour bien conduiré sa raison et chercher la vérité dans les sciences.*

Coordinate geometry may be thought of as the marriage of algebra and geometry. It was a pivotal step in the development of calculus and greatly accelerated the new role of mathematics in the service of science. In fact, John Stuart Mill called it "the greatest single step ever made in the progress of the exact sciences."

---

[10] Actually, Fermat invented this notion of analytic geometry but used an older style notation causing Descartes to get the credit.

We shall use the style and terminology of modern times to help the reader follow the discussion. The idea starts quite simply by constructing two axes, one horizontal and the other vertical. On these axes, usually called the x-axis and y-axis, we have all the numbers (negative, positive, and zero). The place where the axes cross is called the *origin* and is the zero point of each line. Figure 9-1 shows this and plots several points. The x-coordinate is always plotted first. To locate P from the origin, one would go forward 2 units, and then go up 3. Getting to Q from the origin entails going backward 2 and then rising 2. The coordinates of the origin are (0,0), of course.

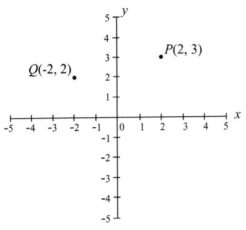

Figure 9-1

Now the beautiful thing that Descartes noticed was that if one were to plot all of the points whose coordinates satisfy an equation of the sort $y = mx + b$, where m and b are constant numbers, one would obtain a straight line! Furthermore, the number b is the y-coordinate of the *y-intercept* – the point where the line crosses the y-axis. The number m is the slope of the line. The slope measures the steepness of the line. It is obtained by knowing the coordinates of any two points on the line, say, (x_1, y_1) and (x_2, y_2), and then taking the ratio

$$\frac{y_2 - y_1}{x_2 - x_1}$$

This is often written as

$$\frac{\Delta y}{\Delta x}$$

The Greek letter delta means *difference*, or *gain*. The points P and Q in Figure 9-2 determine a line segment PQ whose slope is $\frac{1}{3}$. This is because $\Delta y = 3 - 1 = 2$, and $\Delta x = 7 - 1 = 6$. The slope of PQ is the ratio $\frac{2}{6}$ which reduces to $\frac{1}{3}$. Notice the point R directly below P and level with Q. As a consequence, it has x-coordinate 7 and y-coordinate 1. With a bit of luck, it is easy to see that the vertical line RP has length 2 and the horizontal line QR has length 6. Thus, Δy and Δx yield the lengths of the legs of the right triangle associated with any line segment. This made it easy for Descartes to find the lengths of line segments in the so-called *Cartesian plane*. He applied the Pythagorean Theorem to the right triangle. In this example, we get

$$\sqrt{6^2 + 2^2} = \sqrt{40}$$

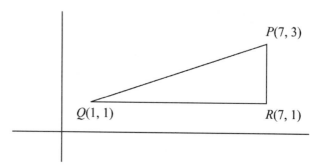

Figure 9-2

More generally, if we denote slope by m and length by L, we get the following formulas for the slope and length of a line segment:

$$m = \frac{\Delta y}{\Delta x}$$

$$L = \sqrt{\Delta x^2 + \Delta y^2}$$

By taking one point of a given line as the y-intercept $(0, b)$ and another point as the generic (x, y), Descartes got the equation

$$m = \frac{y - b}{x - 0}$$

using the above formula for slope. This becomes

$$m = \frac{y - b}{x}$$

Then multiply both sides by x, yielding $mx = y - b$ or, equivalently, $y - b = mx$. Finally, we add b to both sides, obtaining the famous formula $y = mx + b$, taught to all high school students across the world (and possibly on other planets!)

The really big surprise Descartes got was after he plotted many points of $y = x^2$, shown in Figure 9-3. He obtained the parabola of Apollonius, the very same parabola which Galileo found to be the (upside-down) trajectory of a thrown object! Similar equations, such as $y = ax^2 + bx + c$ also yield parabolas. They differ in width and location.

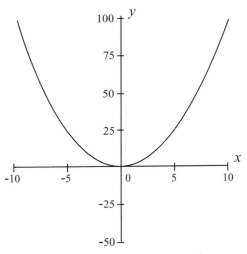

Figure 9-3

By the way, Descartes is known for his quote, "I think therefore I am."[11]

Scientists use the Cartesian plane every day to *graph* the relationship between two variable quantities, such as time versus annual rainfall or pressure versus temperature in a chemical reaction. Graphing relationships helps us state them in mathematical terms and analyze their properties. We often use the expression $f(x)$ (meaning literally "a function of x") to denote a mathematical expression involving x, such as $10x^3 + 3x^2 - 1$, or $\dfrac{5\sqrt{x}}{x^2 - 2x}$. This permits us to

---

[11]Legend has it that he seated himself in a famous Parisian restaurant (perhaps the Binomial Café?) and the waiter asked him if he wanted a drink before dinner. He said "I think not" and promptly disappeared.

write, abstractly, $y = f(x)$, when y depends on x. Of course, one can use letters other than x and y to denote a functional relationship between variable quantities such as pressure and temperature. One would then label the horizontal axis t for temperature, and the vertical axis p for pressure. The relationship could be written abstractly as $p = f(t)$, where f represents the mathematical actions which one must do to t in order to obtain p.

To summarize, René Descartes showed that

1. geometric curves can be described by algebraic equations, and

2. algebraic relationships given by $y = f(x)$ can be graphed in the Cartesian plane to yield a picture of the relationship.

Let's demonstrate this with an example.

Example 9-4 *Graph the straight line* $y = 3x + 1$.

The simplest way to do this is to plot points by choosing appropriate values for x and determining the y value from the equation. Thus, we complete the following table:

x	y
-1	-2
0	1
1	4

Then we plot the points on the Cartesian plane and connect them with a straight line, giving us the desired graph shown below.

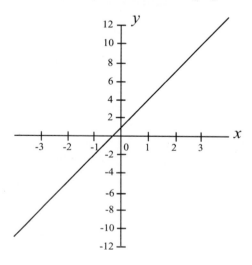

Perhaps the most important hero of the seventeenth century, Isaac Newton[12] (1642-1727), made important contributions to physics as well as mathematics, setting a trend for the next several hundred years. It was already obvious to the thinkers of that century that mathematics was the language of science. Kepler and Galileo used mathematics to express scientific laws with amazing beauty and accuracy. Newton continued this tradition, founding the calculus and establishing the basic laws of physics, still in use today, though they have been augmented and modified by the theory of relativity in the twentieth century. Isaac Newton was a humble man who acknowledged that he stood "on the shoulders of giants." Although Newton showed no particular promise in his elementary education, his genius emerged when he was in his twenties. At the young age of 27, he became a professor at Cambridge after his teacher Isaac Barrow resigned so that his talented student could have his chair.

It is sadly ironic or at least coincidental that Newton was born in 1642 – the year in which one of his greatest "giants," Galileo, died.

The calculus begins with a seemingly easy question. Given a point on a curve in the xy-plane, how does one find the slope of its tangent line? This is another way of asking how steep a curve is at some particular point. While a straight line maintains the same steepness throughout, measured by its slope, a curve has variable steepness. Furthermore, the calculation of a slope requires two points, while we are given only one point here. Newton solved this problem by considering two points: the given point, say P, with coordinates (x, y) and a nearby point on the curve, say Q, with coordinates $(x + \Delta x, y + \Delta y)$. The slope of the line segment PQ is now easily calculated to be $\dfrac{\Delta y}{\Delta x}$. This is, unfortunately the slope of the chord PQ, and not the slope of the desired tangent. Now consider what happens as the point Q slides down the curve toward P. While the quantities Δx and Δy grow smaller, their ratio approaches the desired slope! This process is called "taking the limit."

---

[12]There was quite a controversy as to just who exactly invented calculus, Newton or Leibniz. What is clear is that each man developed his own version of calculus independently of the other. Newton was first by about a decade but Leibniz was the first to publish his ideas. They are both considered to be the fathers of calculus. It should also be mentioned that more than a decade before either man was born, Fermat had developed many of the techniques used in calculus today.

Let's illustrate this for the parabola $y = x^2$, pictured in Figure 9-4.

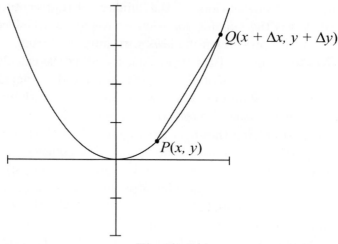

Figure 9-4

The point P has coordinates (x, y) and Q has coordinates $(x + \Delta x, y + \Delta y)$. Now let's do some fearless calculating. If $y = x^2$, then it stands to reason that

$$y + \Delta y = (x + \Delta x)^2 = x^2 + 2x\,(\Delta x) + (\Delta x)^2$$

(keeping in mind Δx is one creature) from which it follows by subtraction of the equation $y = x^2$ that

$$\Delta y = 2x\,(\Delta x) + (\Delta x)^2$$

If we divide both sides of this last expression by Δx, we get the slope of chord PQ:

$$\text{Slope of } PQ = \frac{\Delta y}{\Delta x} = 2x + \Delta x$$

Now if we want Q to approach P, this is the same as requiring Δx to approach zero. We do this by letting Δx be zero in the last expression, yielding the answer!! The slope of the tangent to the parabola $y = x^2$ at the point (x, y) is given by the new expression $2x$, derived from the original expression by this new limiting process applied to the ratio of differences $\dfrac{\Delta y}{\Delta x}$. This process is called *differentiation* and the new expression which yields tangent slopes for different values of x is called the *derivative*, denoted by the symbol y', to distinguish it from the original expression for y which yields height. The notation $f'(x)$ is also commonly used to distinguish the derivative from the original function $f(x)$.

Let's reconsider the above problem from a slightly different point of view. Let x represent time (in hours) and let y represent distance (in miles) and let the equation $y = x^2$ yield the relationship between time of travel (starting at noon, represented by $x = 0$) and the distance covered in that time. At noon, $x = 0$ and $y = 0$, that is, our exciting trip is just starting. At one o'clock, $x = 1$ and $y = 1$. We are obviously on a highway during the rush hour as we have traveled one mile in a whole hour. Now at three o'clock, $x = 3$ and $y = 9$, so our average speed in the first three hours of travel is, using the famous equation

$$rate = \frac{distance}{time}$$

9 miles/3 hours $= 3$ miles per hour. Wow, we hope your seatbelt is on. What about the average speed from 1 P.M. to 3 P.M.? Well $y = 3^2 - 1^2 = 9 - 1 = 8$ miles, while $x = 3 - 1 = 2$ hours, so the average speed is 8 miles/2 hours $= 4$ miles per hour. The point is that average speed is given by the slope, that is, $\frac{\Delta y}{\Delta x}$, of the chord! Then the tangent slope given by the derivative must be the *instantaneous velocity*! The expression $2x$, obtained above yields the instantaneous velocity at time x. At 10 P.M., for example, the speedometer actually says 20 mph.

This is very important in physics. The momentum of a speeding truck is determined by its instantaneous velocity. The average speed of a car never hurt anybody! Speeding tickets are not given for excessive average velocities.

Newton (and others) showed that the method of differentiation could be extended to other expressions. If $y = ax^n$, he showed that the derivative is given by

$$y' = nax^{n-1}$$

Thus if an object falls y feet in x seconds, then $y = 16x^2$, as Galileo showed; then its velocity is given by $y' = 32x$, in perfect agreement with Galileo's observation that the gravitational acceleration is 32 ft/sec each second. Thus, if $x = 5$, the velocity is $32 \times 5 = 160$ ft/sec because the velocity increases by 32 ft/sec each second.

The second part of the calculus reverses the process of differentiation. Most mathematical operations can be undone and differentiation is no exception. Newton discovered that the area above the x-axis and under a given curve, between the vertical lines $x = a$ and $x = b$, can be computed by finding the *antiderivative* and subtracting its value at $x = a$ from its value at $x = b$. The *antiderivative*, sometimes called the *integral*, of a function $f(x)$ is a new function $F(x)$ whose derivative is $f(x)$.

Now since the derivative of

$$ax^n$$

is

$$nax^{n-1}$$

which in words says, "multiply the coefficient by the power of x and then lower this power by 1," we may conclude that the antiderivative of

$$ax^n$$

is

$$\frac{ax^{n+1}}{n+1}$$

that is, raise the power by 1 and divide by the new power. (Socks and shoes, remember?)

Let us find the area of the region R of Figure 9-5, bounded by the x-axis, the vertical lines $x = 1$, $x = 2$, and the parabola $y = x^2$. Observe that the a and b of this problem are 1 and 2, that is, we start at $x = 1$ and end at $x = 2$. Since the antiderivative of x^2 is

$$\frac{x^3}{3}$$

we compute the difference

$$\frac{2^3}{3} - \frac{1^3}{3} = \frac{7}{3}$$

That is the exact area under the parabola! After five thousand years of struggle, the general area problem was finally solved!

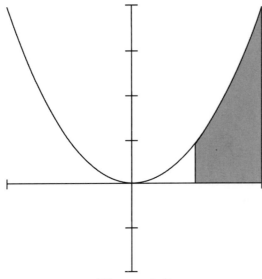

Figure 9-5

Newton discovered several important laws of physics – the science of matter, energy, and motion. His law of inertia states that a body moving in a straight line with constant velocity will continue doing so unless it is acted on by a force. This also implies that a body at rest will continue to do so unless acted on by a force. Rest is a state of constant velocity – namely zero!

This law is hard to deduce here on earth because of gravity. In outer space, a moving object will keep on moving. Once a rocket leaves the gravitational field of the earth and is aimed correctly, the engine may be shut off without any loss in speed. Of course, force must be applied to alter the rocket's course, as Newton correctly predicted.

His famous equation $F = ma$ says that the force applied to an object is the product of the mass and the acceleration. *Acceleration* is defined as the change in velocity. It is a consequence of a force applied to the object of mass m. If we rewrite this equation as

$$a = \frac{F}{m}$$

we see that the same force applied to two objects of differing masses will accelerate the less massive object more, since a smaller denominator produces a larger fraction, for example, $\frac{5}{6}$ is greater than $\frac{5}{8}$.

Newton took the work of Galileo and Kepler (and others) on planetary motion and, assuming that the sun's gravitational pull

held the solar system together, developed the law of gravity which
states that any two bodies attract each other with a force F given
by the formula

$$F = \frac{gMm}{d^2}$$

in which M and m represent the masses of the two bodies, g is a
gravitational constant, and d^2 is the square of the distance between
them. Newton showed that the elliptic orbits of the planets are a
consequence of this formula, as are Galileo's results for objects falling
under the influence of gravity.

Three observations are in order here.

1. The role of mathematics as the servant of science was firmly
   established by the close of the seventeenth century.

2. Newton established a cold, mechanistic universe consisting of
   matter in motion, perfectly described by mathematical laws.

3. The future can be predicted by the current state of the universe!
   Tell Newton an object's location, velocity, and the forces acting
   on it, and he will be able to determine its future trajectory and
   speed.

This had a profound effect on the intellectual climate of Europe. Philosophically, it led to the Enlightenment, or as it is sometimes called, the Age of Reason. Thinkers assumed that the universe was open to reason and discovery. This gave science, technology, and enterprise a motivating jolt. It wasn't long before the telescopes, microscopes, laboratories, scientific journals, machines, and, of course, the brains of Europe got very busy in pursuit of remarkable achievements in a host of fields. No one was immune from the frenzy. Even political theorists such as John Locke started invoking laws of nature to justify political and social theories which had profound consequences in the affairs of nations.

It is not a coincidence that the starting date of the Enlightenment is designated by historians as 1687, the year Isaac Newton published his magnum opus, *Principia Mathematica*. Newton was inducted into the Royal Society in 1672. He was appointed Minister of the Mint in 1699 and knighted by the queen in 1705.

Partially as a result of the work of Newton and other mathematicians of the seventeenth century, the next one hundred years were characterized by an optimistic worldview, typified by Voltaire, Benjamin Franklin, and the founders of the United States of America who dared to challenge the authority of God, King, and Empire. (The Enlightenment ends, by the way, with the start of the French Revolution.)

The scientists of the eighteenth century capitalized on the mathematical advances of the seventeenth to produce an incredible body of theoretical knowledge and its applications that would eventually bring about the Industrial Revolution and the amazing technological innovations of the modern age. Little did the mathematicians of the exciting seventeenth century imagine how far their discoveries propelled us forward in the never-ending struggle of science to harness the principles of our natural universe for the betterment of humankind.

1. Marin Mersenne,[13] a Franciscan friar, philosopher, scientist, and mathematician, asked Fermat if the number 100,895,598,169 is prime. Fermat wrote back saying that it was the product of 112,303 and 898,423, and hence not prime. Show that Fermat was correct by multiplying these two numbers. Fermat also stated that these two factors were prime. (To this day no one knows how Fermat did it.) Can you show that these factors are prime?

2. Show that Fermat's Little Theorem

   $$a^{p-1} \equiv 1 \, (\mathrm{mod}\, p)$$

   is true for each of the following cases:
   (a) $p = 3$ and $a = 5$ (b) $p = 3$ and $a = 7$
   (c) $p = 5$ and $a = 2$ (d) $p = 5$ and $a = 4$

3. Make a chart of the numbers from 0 to 100 modulo 6. Why does the first row consist of the numbers 0, 1, 2, 3, 4, and 5? If the numbers in the first column are all of the form $6k$, how can you characterize the numbers in each of the other columns?

4. True or false:
   (a) $70 \equiv 30 (\mathrm{mod}\, 10)$ (b) $35 \equiv 24 (\mathrm{mod}\, 5)$
   (c) $13 \equiv 19 (\mathrm{mod}\, 2)$ (d) $30 \equiv 14 (\mathrm{mod}\, 4)$
   (e) $37 \equiv 13 (\mathrm{mod}\, 11)$ (f) $-1 \equiv 2 (\mathrm{mod}\, 3)$

---

[13]He is often referred to as "the father of acoustics" for his work in that field.

5. Verify the following equations involving the first few powers of 2: $2 \equiv 2 \pmod 3$, $4 \equiv 1 \pmod 3$, $8 \equiv 2 \pmod 3$, $16 \equiv 1 \pmod 3$, $32 \equiv 2 \pmod 3$. Based on these values, predict the mod 3 values of the next few powers of 2. Now prove that this pattern continues for all powers of 2, using Exercise 4(f). (*Hint:* $(-1)^n$ is $+1$ when n is even and is -1 when n is odd.)

6. Prove the six laws of modular arithmetic stated in this chapter.

7. Convert the following to yards, feet, and inches:

   (a) 300 inches
   (b) 150 inches
   (c) 178 inches
   (d) 250 inches
   (e) 700 inches
   (f) 1000 inches

8. Verify that the first few squares,

$$1, 4, 9, 16, 25, 36, 49, 64, 81, \text{ and } 100$$

   follow the following pattern mod 3:

$$1, 1, 0, 1, 1, 0, 1, 1, 0, 1, \ldots$$

   In other words, every third square equals $0 \pmod 3$, while the rest equal $1 \pmod 3$. Notice that no square seems to equal $2 \pmod 3$. Prove this using Fermat's Little Theorem. (*Hint:* Let $p = 3$, in which case $p - 1 = 2$.)

9. If it is now 3 P.M., what time will it be in 123 hours? Explain the relevance of modular arithmetic to this problem. Which modulus should be used? Why is 12 not the best modulus when using military time?

10. If you are trying to convert inches into feet and inches (e.g., 30 inches = 2 feet and 6 inches), how is modular arithmetic useful? Make up several additional practical examples involving modular arithmetic.

11. Find the entries of the seventh row of Pascal's triangle using the entries of the sixth row of the chart in this chapter. Show that these entries are the combinatorial numbers

$$C(7,0), C(7,1), C(7,2), \ldots, C(7,7)$$

Verify that the sum of the entries in the seventh row is 2^7.

12. If one flips a coin three times in a row, what is the probability of getting exactly one head? Do this by listing all possible outcomes and then counting how many of them are favorable. Then use the ratio

$$\frac{\text{favorable outcomes}}{\text{total outcomes}}$$

13. A restaurant menu features 3 salads, 4 soups, 6 main dishes and 2 desserts. In how many ways can a customer order a complete dinner? What is the probability of correctly guessing what the customer ordered?

14. Why is $2x + 3y = 6$ the equation of a straight line? Show how to convert it into the form $y = mx + b$. Since two points determine a straight line, graph the line $2x + 3y = 6$ by selecting two points, plotting them and connecting them with a straight line. A simple way to obtain a point on a line is to substitute an easy value for one of the variables and then solving for the other. 0 is often a convenient guess for either x or y.

15. Graph the following lines:

   (a) $4x + 2y = 10$
   (b) $5x - y = 10$
   (c) $y = 2x + 7$
   (d) $x = 12 - 2y$
   (e) $\dfrac{x}{2} + \dfrac{y}{5} = 1$
   (f) $y = 5$.

16. Since y may be thought of as *height*, what is the relationship between the graphs of $y = x^2$, $y = x^2 + 2$, and $y = x^2 + 5$? Graph these parabolas on the same set of axes.

17. Find the derivatives of the following functions. Assume that the derivative of a sum of several terms is the sum of the derivatives of the terms. Note, also, that the derivative of a constant term is 0. (Explain why this is true.) Why is the derivative of x equal to 1?

(a) $y = 6x^2 + 6x + 6$

(b) $y = 3x^3 - 5x^2 + 2x - 8$

(c) $y = 2x^4 - 6x^3 + 10x^2 - 4x$

(d) $y = mx + b$ (m and b are constants)

18. Find the area under the parabola $y = 3x^2$ between the vertical lines $x = 1$ and $x = 3$ by finding the antiderivative and subtracting its value at $x = 1$ from its value at $x = 3$.

19. **Zeller's Congruence.** *Zeller's congruence* is a formula for determining the day of the week for any given date. The formula is

$$weekday = \left(D + \left[\frac{13M - 1}{5}\right] + Y + \left[\frac{X}{4}\right] + \left[\frac{Y}{4}\right] - 2X\right) \bmod 7$$

where D is the day of the month, X is the first two digits of the year, Y is the last two digits of the year, and M is the month according to the numbering, $March = 1$, $April = 2$, $May = 3$, $June = 4$, $July = 5$, $Aug. = 6$, $Sept. = 7$, $Oct. = 8$, $Nov. = 9$, $Dec. = 10$, $Jan. = 11$, and $Feb. = 12$. Because of this wacky numbering, it is important to subtract 1 from the year when dealing with a date in January or February. For example, the date of Jan. 7, 1999 gives us $M = 11$, $D = 7$, $X = 19$ and, after subtracting 1, $Y = 98$. Plugging this into the formula, gives us

$$weekday = \left(7 + \left[\frac{13 \times 11 - 1}{5}\right] + 98 + \left[\frac{19}{4}\right] + \left[\frac{98}{4}\right] - 2 \times 19\right) \bmod 7$$

$$= \left(7 + \left[\frac{143 - 1}{5}\right] + 98 + 4 + 24 - 38\right) \bmod 7$$

$$= (7 + 28 + 98 + 4 + 24 - 38) \bmod 7$$

$$= 123 \bmod 7$$

$$= 4$$

From the code *Sun.* = 0, *Mon.* = 1, *Tues.* = 2, *Wed.* = 3, *Thurs.* = 4, *Fri.* = 5, and *Sat.* = 6, we see that Jan. 7, 1999 was on a Thursday. Now, try this for the following dates:

(a) Nov. 23, 2000

(b) July 4, 2001

(c) Feb. 17, 2000

(d) Jan. 1, 1999

20. The first four notes of the major pentatonic scale are "do," "re," "mi," and "sol." Write down all 24 permutations of these four notes. How many permutations are possible using the twelve notes of the chromatic scale?

Suggestions for Further Reading

1. Beiler Albert. *Recreations in the Theory of Numbers*. Dover, New York, 1964.

2. Berlinski, David. *A Tour of Calculus*. Vintage, New York, 1995.

3. Descartes, René. *Discourse on Method and the Meditations*. Penguin, New York, 1998.

4. Friedberg, Richard. *An Adventurer's Guide to Number Theory*. Dover, New York, 1994.

5. Hardy, G. H., and Wright E. M. *An Introduction to the Theory of Numbers*. Oxford University Press, New York, 1980.

6. Simmons, George F. *Calculus Gems: Brief Lives and Memorable Mathematics*. McGraw-Hill, New York, 1992.

7. Singh, Simon, and Lynch, John. *Fermat's Enigma: The Epic Quest to Solve the World's Greatest Mathematical Problem*. Bantam Books, New York, 1998.

8. Steen, Lynn A. *On the Shoulders of Giants: New Approaches to Numeracy*. National Academy Press, Washington DC, 1990.

Chapter 10

The Age of Euler

Rarely has the world seen a mathematician as prolific as the great Leonhard Euler[1] (1707-1783). Born in Switzerland, he even-

---

[1] Euler was the person who gave us the notation π for pi, i for $\sqrt{-1}$, Δy for the change in y, $f(x)$ for a function, and $\sum$ for summation.

tually obtained royal appointments in two European courts, namely Russia and Germany (under Frederick the Great). He published so many mathematics articles that his work fills seventy thick volumes. His publications account for one-third of all the technical articles of eighteenth-century Europe. The preceding century saw the rise of scientific and mathematical journals – the new media of the times and the quickest way of making innovations known to colleagues across the continent. This outgrowth of the printing revolution of the fifteenth century accelerated the pace of mathematical and scientific progress by transmitting new ideas in a timely manner – much like the present computer revolution has just begun to affect the dissemination of knowledge.

After Euler's death, it took forty years for the backlog of his work to appear in print. Although he lost his sight in 1768, for the last fifteen years of his life he continued his research at his usual energetic pace while his students copied his pearls of wisdom. It is inconceivable to most how he did mathematics without pencil and paper – without being able to see the multitude of diagrams, equations, and graphs needed to do research.

What areas of math did he enrich and expand? The question is what field of math did he not enrich and expand! Not only did he contribute substantially to calculus, geometry, algebra, and number theory, he also invented several fields. Though a father to eleven children, Euler found time to become the father of an important branch of mathematics, known today as *graph theory*, which would be important in modern fields such as computer science and operations research, as well as traditional areas such as physics and chemistry.

Euler became the father of graph theory as well as topology after solving the notorious "Seven Bridges of Königsberg" problem. The diagram of Figure 10-1 shows the four landmasses of the city of Königsberg and the seven bridges interconnecting them.

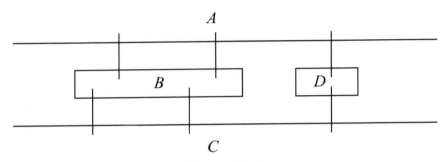

Figure 10-1

The problem was to devise a route that traverses each bridge exactly once and to end where one starts. Euler observed that the task could not be done!! He noticed that each landmass has an odd number of bridges connecting it with the rest of the city. Hence a traveler departing, returning, departing, and so forth, an odd number of times would wind up departing on the last bridge, rendering impossible his return to his point of origin.

Let's consider this gem of thinking one more time. Number the bridges contiguous with landmass A, 1, 2, and 3. Then if one starts the trip by departing A on bridge number one, he must return on bridge number two or number three, leaving only one more bridge. Clearly he must depart on that bridge not yet traveled on – and that makes all the difference! He cannot end his trip on landmass A.

Euler observed that the sizes of the land masses as well as the lengths and shapes of the bridges were irrelevant. Consider, therefore, a diagram representing the landmasses as dots and the bridges as lines, as in Figure 10-2.

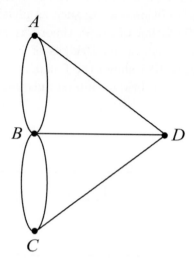

Figure 10-2

Notice the irrelevance of the weird shapes of the bridges meeting at B. The lengths of the lines are, likewise, unimportant. For that matter, so are the precise locations of the dots labeled A, B, C, and D.

In the spirit of Euler, a *graph* is defined as follows. A graph G is a collection of dots (more commonly called *vertices*, as we shall call them from now on), and a collection of lines (called *edges*), each line rendering a pair of vertices *adjacent*, that is, the edge links the two vertices. The specific layout, or representation, of the graph doesn't matter, as long as the adjacencies and nonadjacencies are preserved.

Imagine an airline graph in which London, Paris, and New York City are vertices and the edges between them represent direct flights on Pack'em-In Airways. The issue is simply a yes-or-no question. Which cities are connected by flights? The graph of Figure 10-3 answers this exciting question. Can you guess what N, L and P represent? Are we concerned with the $\angle NLP$, that is, the angle made by edges NL and LP? Of course not. The edges could just as well be curved.

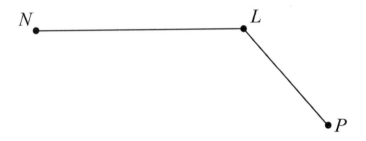

Figure 10-3

Incidentally, in this book we shall not consider graphs in which a single pair of vertices may be linked by more than one edge, as in the graph of the seven bridges problem, where vertices A and B are linked by two edges. Such graphs are today called *multigraphs* and are important in certain transportation problems, for example, in which several airlines fly between various pairs of cities.

The graph G of Figure 10-4 will be used to illustrate several concepts in graph theory. You may interpret this graph any way you wish. Some will think of the vertices as cities and the edges as flights. Others may view the vertices as atoms in a molecule. The edges will then presumably represent bonds between some of the atoms. The vertices may, in a more animated manner, represent members of the board of directors of a company! An edge between two vertices might indicate that they work well together. In fact entire books have been written on the applications of graph theory to a host of different situations. Ah ... before we forget, here is the graph.

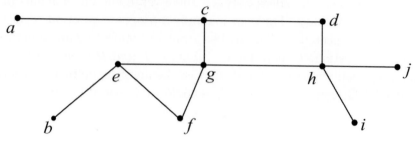

Figure 10-4

The *degree* of a vertex is the number of edges touching it (technically, *incident* with it). Thus the degree of vertex g in graph G above is 4. This can be written compactly as $\deg(g) = 4$. Graphs are usually identified by capital letters and the vertices are denoted by lowercase letters. Edges may also be labeled using small letters, but the common practice is to label an edge using the letters of the two vertices it is incident with. The rightmost edge in graph G above, for example, may be referred to as edge hj.

The set of vertices and the set of edges of a graph G are denoted $V(G)$ and $E(G)$, respectively. Many authors adhere to the convention that n and e represent the cardinalities (i.e., sizes) of the vertex set and edge set, respectively. In the above example,

$$V(G) = \{a, b, c, d, e, f, g, h, i, j\}$$

in which case $n = 10$, that is, graph G has ten vertices. You should be able to verify that

$$E(G) = \{ac, be, cd, cg, dh, ef, eg, fg, gh, hi, hj\}$$

implying that $e = 11$, that is, G has eleven edges. Vertices a, b, i, and j have degree 1 and are therefore called *endvertices*.

Euler established the following interesting fact, important enough to be called a theorem.

Theorem 10-1 *The sum of the degrees of the vertices of a graph equals twice the number of edges.*

Proof. The proof is easy! Each edge contributes one to each of the degrees of the two vertices to which it is adjacent. Hence the degree sum is twice the number of edges. ∎

As a consequence, (and mathematicians are always on the lookout for consequences) the sum of the degrees of any graph must be an even number. Interesting.

Some graphs can be categorized into classes. For example, the graph in Figure 10-5, called C_4, belongs to the important class of graphs known as *cycles*. If a cycle has n vertices, we call it C_n. Note that each vertex of a cycle has degree 2. A graph (not necessarily a cycle) in which each vertex has the same degree is called *regular*. If the common degree is r, we call the graph *r-regular*. Thus, for example, the cycles C_n are 2-regular.

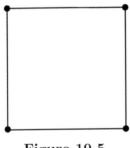

Figure 10-5

Remember to attach no significance to the shape of the cycle. The edges can have different lengths and can even be curved. Of course, most graph theorists will attempt to draw the graph so that our attention is drawn to its symmetries or other important properties.

Another very important class of graphs are *paths*, denoted P_n, where n is, once again, the number of vertices in the path. Figure 10-6 illustrates P_5.

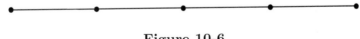

Figure 10-6

In this short account of graph theory, we will assume that the graphs we study are *connected*, that is, there is a way to travel between any two vertices by traversing a sequence of consecutive edges between them. In the graph G of Figure 10-4, for example, one can travel from vertex b to vertex d by traversing the consecutive edge sequence be, eg, gc, cd. Another way to say this is that there is a path in the graph whose end points are b and d. This is called a *b-d path.* The vertices of this path form a sequence in which consecutive members are adjacent. Note that there is another *b-d* path with vertices *b*, *e*, *g*, *h*, and *d*. This is useful information if the graph is an airline graph and the airport in city *c* is closed. We can then reroute the traveler from city *b* to city *d* by flying from *g* to *h* instead of from *g* to *c*. The same logic would apply if *c* were a telephone exchange that is malfunctioning.

The reason we have travel options is that graph *G* contains cycles, namely C_3, with vertices *e*, *f*, and *g*, and C_4, with vertices *c*, *d*, *g* and *h*. A connected graph without cycles is called a *tree*. It is characterized by the fact that there is only one path between any pair of vertices. A path is an example of a tree. The graph *T* of Figure 10-7 shows a more general tree on ten vertices of varying degrees. Note that it has nine edges. This is not a coincidence.

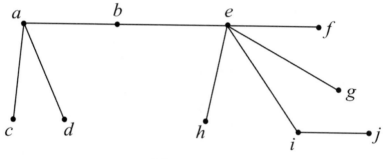

Figure 10-7

A tree always satisfies the equation $e = n - 1$, that is, the number of vertices is one greater than the number of edges. This can be proven by drawing the tree one vertex and one edge at a time! At stage one, we will have a single vertex with no edges. Thus $n = 1$ and $e = 0$, exactly as the tree equation predicts. Now draw an edge and an adjacent vertex. We now have $n = 2$ and $e = 1$, again satisfying the tree equation. Now attach a third vertex (and edge) to the drawing, making $n = 3$ and $e = 2$. Get the idea? At each step, we add one edge and one vertex, raising both *n* and *e* by one but not altering the fact that the vertices are always ahead of the edges by one.

It follows, by the way, that if a class full of attentive students is told to draw any tree they wish containing exactly ten vertices, they will draw many different kinds of trees but they will all have exactly nine edges! A rebellious student who uses only eight edges will not get a connected structure. (An equally rebellious student who uses ten edges will get a cycle.) For this reason, a tree is often called *minimally connected*. It is precisely the absence of cycles that accounts for this important property. An example of a tree we use in daily life is the family tree – which, hopefully, contains no cycles. The paths P_n form an important subclass of trees. We use them to measure distances in graphs between pairs of vertices.

We define the *distance* between vertices x and y of graph G to be the length of a shortest x-y path. We denote it by $d_G(x, y)$. The subscript G may be omitted if the context is clear. Notice that we didn't say "the length of *the* shortest x-y path," since there may be more than one shortest path. In the graph G of Figure 10-4, for example, in which $d(g, d) = 2$, there are two g-d paths of length 2, that is, consisting of 2 edges.

In a tree, there is only one path between any pair of vertices.

This has drawbacks as well as advantages. In a communications network modeled after a tree, if one edge fails, the entire network is disconnected and we have two components which cannot communicate with one another. Cycles give us choices – ways to get around a failed edge (or vertex). Bear in mind that the Internet is a giant graph in which each vertex is a computer. The World Wide Web could be renamed the *world wide graph* – but rest assured that this is very unlikely.

In the nineteenth century, chemists began using graphs to predict configurations of saturated hydrocarbons. (Stay with us here – don't close the book!) A carbon atom has valence 4, meaning that it can form four single bonds with other atoms. Hydrogen, the most abundant and simplest atom in the universe, has valence 1. Thus a single carbon atom can bond with four hydrogen atoms to form the molecule called *methane*. Chemists denote this molecule CH_4. We call this hydrocarbon *saturated* because the carbon's four bonds are all in use.

What would a saturated hydrocarbon look like if it had two carbons? They could bond with each other leaving each free to attract three hydrogens yielding the molecule C_2H_6. This is called *ethane*. In a similar manner, imagine a path of three carbons. The one in the middle can attract only two hydrogens, since two of its bonds are with the other carbons, which, by the way, can attract three hydrogens apiece, bringing the total number of hydrogens to eight. This is a *propane* molecule, denoted C_3H_8.

So far, so good. The fun starts when we consider four carbon atoms. They can form a path or a so-called *star* in which a central carbon bonds with the other three. In each case, however, they are saturated when they attract a total of ten hydrogen atoms. We obtain two different molecules though they are both denoted C_4H_{10}. The first is called *butane* (as in cigarette lighters) and the other is called *isobutane*. The two molecules are referred to as *geometric isomers*. The point is that graph theory predicts these possible configurations. The two isomers are shown in Figure 10-8. Warning: This diagram is highly flammable!

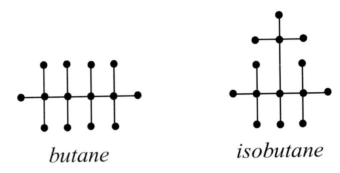

butane   *isobutane*

Figure 10-8

Given a graph G and a vertex x, we define the *eccentricity* of x, denoted $e(x)$, to be the distance of x from a farthest vertex in G. One may think of the eccentricity of x as its "worst case scenario" in regard to its distances to all the vertices of G. To illustrate this concept, we label the eccentricities of the vertices of the path P_7 in Figure 10-9.

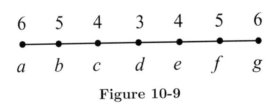

Figure 10-9

Notice that $e(c) = 4$ because $d(c,g) = 4$ and this is the maximum distance of vertex c from any of the vertices of the path. Observe that while $e(d) = 3$, there are two vertices at distance three from vertex d, namely a and g.

The *radius* and *diameter* of a graph G [denoted $rad(G)$ and $diam(G)$] are defined to be the smallest and largest eccentricity in the graph, respectively. The path of Figure 10-9 has a radius of 3 and a diameter of 6. The set of vertices with minimum eccentricity is called the *center* of G and is denoted $C(G)$. The center of the path in Figure 10-9 consists of the lone vertex d.

The exercises at the end of this chapter will enhance your understanding of the material on graph theory while keeping you thoroughly entertained. Many mathematicians are motivated by their love of the subject – and the joy of discovery. The applications to science, technology, and so forth, are often seen as mere consequences. To some, math is like a game of chess or a crossword puzzle. To others, it is a noble crusade in the religious pursuit of truth for the betterment of humanity. To some teachers many of us have been plagued with, it's just a job. To the great Euler, it was a raison d'être.

Example 10-1 *For the graph below, list the vertices and edges, and find the degree of each vertex.*

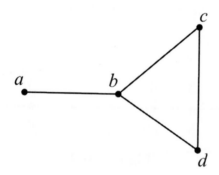

The vertex set is $V(G) = \{a, b, c, d\}$, the edge set is $E(G) = \{ab, bc, bd, cd\}$, and $\deg(a) = 1$, $\deg(b) = 3$, $\deg(c) = 2$, and $\deg(d) = 2$. ∎

Euler did a great deal of work in number theory. He did pioneer work in the theory of partitions. A partition of a given integer is simply a way of writing it as a sum of integers. The partitions of 7 are, for example,

$$7$$
$$6 + 1$$
$$5 + 2$$
$$5 + 1 + 1$$
$$4 + 3$$
$$4 + 2 + 1$$
$$4 + 1 + 1 + 1$$
$$3 + 3 + 1$$
$$3 + 2 + 2$$
$$3 + 2 + 1 + 1$$
$$3 + 1 + 1 + 1 + 1$$
$$2 + 2 + 2 + 1$$
$$2 + 2 + 1 + 1 + 1$$
$$2 + 1 + 1 + 1 + 1 + 1$$
$$1 + 1 + 1 + 1 + 1 + 1 + 1$$

Notice how many there are. Imagine how many partitions there must be for a number like 5,167,855,984,750. Sticking to our modest example of 7, Euler chose another smaller number, say 3, and then counted two different subclasses of partitions of 7:

(a) Those with 3 or fewer terms, such as $4 + 3$.

(b) Those with terms less than or equal to 3, such as $2 + 2 + 2 + 1$.

These classes are not mutually exclusive. The partition $3 + 3 + 1$ satisfies both conditions. To his surprise, both groups have the same number of partitions!! Euler showed that this is always the case. He proved the following theorem.

Theorem 10-2 *Given integers n and k, such that $k < n$, the number of partitions of n containing no more than k terms equals the number of partitions containing no term greater than k.*

He proved it by setting up a one-to-one correspondence between members of both groups of partitions as follows. Represent a partition by a collection of dots by forming a matrix whose rows of dots represent the terms of the partition. Let's illustrate this for the partition in the first group of our example above given by requirement (a), namely $4 + 3$. We get

$$\begin{matrix} \bullet & \bullet & \bullet & \bullet \\ \bullet & \bullet & \bullet & \end{matrix}$$

Euler matched this partition with its so-called *transpose*, that is, the matrix formed by turning the first row into the first column, the second row into the second column, and so on, obtaining

$$\begin{matrix} \bullet & \bullet \\ \bullet & \bullet \\ \bullet & \bullet \\ \bullet & \end{matrix}$$

This establishes a correspondence with the partition $2 + 2 + 2 + 1$ which belongs to the second group given by requirement (b). The reason he knew the transpose of a graph of a partition of the first group would be the graph of a partition in the second group is that the graph would have at most three rows, that is, its columns have at most three dots. It follows that the transpose has rows with at most three dots, that is, its terms are no larger than 3. This is an incredibly clever argument.

Let's look at another simpler example.

Example 10-2 *List the partitions of 4.*

When partitioning a number n, we obtain from 1 to n terms. So we can have from 1 to 4 terms. With one term, we get 4. With two terms, we have $3 + 1$ and $2 + 2$. Partitioning 4 into 3 terms yields $2 + 1 + 1$. Finally, using 4 terms gives us $1 + 1 + 1 + 1$. ∎

Another mathematician in the court of Frederick the Great was a Frenchman, Joseph-Louis Lagrange[2] (1736-1813). He was a modest self-taught scholar who contributed hundreds of papers to the French Academy and the Berlin Academy, in addition to founding his own mathematics journal. He applied calculus to physics and astronomy and showed that in a closed system under the influence

---

[2]Lagrange once said, "When we ask for advice, we seek an accomplice."

of gravity, the sum of the kinetic and potential energy of a moving body remains constant. The kinetic energy of a body with mass m and velocity v is $\frac{1}{2}mv^2$, while its potential energy is mgh, in which g is the gravitational acceleration (first studied by Galileo) and h is its height above the ground.

At the moment of release, the kinetic energy is zero since $v = 0$. When the object hits the ground, the potential energy is zero. Then the initial potential energy must equal the terminal kinetic energy. Upon equating these two expressions but using V and H (instead of v and h) to represent terminal velocity and initial height, he obtained the equation

$$\frac{1}{2}mV^2 = mgH$$

which after canceling the m on both sides, then transposing the $\frac{1}{2}$ on the left to a 2 on the right, and finally square rooting both sides, becomes

$$V = \sqrt{2gH}$$

Since $g = 32$ ft./sec/sec, this becomes

$$V = 8\sqrt{H}$$

If you drop a rock from a height of 100 feet, it will hit the ground with a velocity of $8 \times \sqrt{100}$ or simply 80 ft/sec, neglecting air resistance. Warning: This velocity can be terminal in more than one sense of the word. Do not do this unless you are a construction worker and there is a large, clear area below in which it is safe to throw objects. (We added this warning to be protected from lawsuits.)

Like Euler, Lagrange was interested in number theory and proved that any integer can be represented by the sum of four (or fewer) squares. This is surprising since perfect squares get rarer as the numbers get larger, that is, consecutive perfect squares get further and further apart. After 9, the next square, 16, is 7 units away. On the other hand, after 100, the next square, 121, is 21 units away. Here are some examples of Lagrange's amazing theorem.

$$20 = 16 + 4$$
$$21 = 16 + 4 + 1$$
$$22 = 9 + 9 + 4$$
$$23 = 9 + 9 + 4 + 1$$
$$24 = 16 + 4 + 4$$
$$25 = 16 + 9$$
$$26 = 16 + 9 + 1$$
$$27 = 9 + 9 + 9$$
$$28 = 9 + 9 + 9 + 1$$
$$29 = 25 + 4$$
$$30 = 25 + 4 + 1$$

The eighteenth century saw an extension of Descartes' analytic geometry and Newton's calculus to three dimensions, that is, to space. Imagine three mutually perpendicular axes meeting at a common origin. Two of them are the old familiar x- and y-axes of a horizontal plane, while the newly added z-axis is vertical, denoting height above the xy-plane, or depth below it. Thus, the third coordinate is the z-coordinate and is 0 for points in the xy-plane, positive for points above it, and negative for points below it. Figure 10-10 depicts a point P with coordinates a, b, and c. It is located directly above the point Q in the xy-plane, with coordinates a, b, and 0. The parallelogram in Figure 10-10 is really a rectangle and its intercepts on the x- and y-axes are shown. Note that we see only the positive axes in the figure. One must imagine three right angles at the origin, though only one of the angles looks as if it is right.

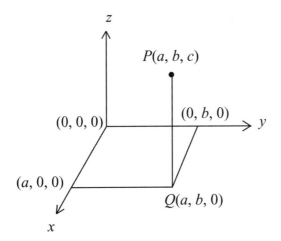

Figure 10-10

Imagine graphing $z = x^2 + y^2$. (See Figure 10-11.) Each point in the xy-plane determines an altitude and, therefore, a point in space above it. For example, directly above the point $(2, 3, 0)$ in the xy-plane, one would graph the point $(2, 3, 13)$ since $2^2 + 3^2 = 4 + 9 = 13$. Neat!! We get a surface instead of just a curve. This enables us to build and analyze three-dimensional structures. But it doesn't end here. We can now analyze phenomena in which a variable quantity such as the pressure of a gas depends on the temperature and the volume of the container (assuming that these quantities vary, too).

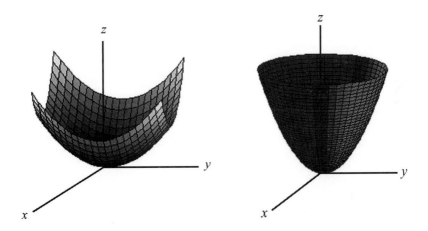

Figure 10-11: Two Views of the Paraboloid $z = x^2 + y^2$

The ability to do this is vital to scientists. Many variables in nature depend on several other variables. It is highly simplistic to be content with $y = f(x)$. In three-dimensional space, we can visualize $z = f(x, y)$, read "z is a function of x and y."

Using x-, y-, z-space, mathematicians are also able to analyze twisting curves, from the trajectories of birds or the more complex trajectories of rapidly moving particles of sugar in a cup of tea to the tiny strands of DNA that determine our genetic makeup. (See Figure 10-12.)

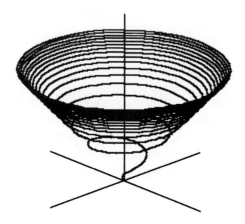

Figure 10-12: A Space Curve We Call Taz

In summation, continuing the trend established in the previous century, the eighteenth century saw a great leap in both the quality and quantity of mathematical knowledge and power.

EXERCISES

1. For each of the following graphs, list the vertices and edges and find the degree of each vertex.

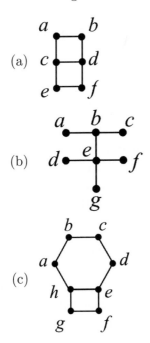

(a)

(b)

(c)

2. For each of the following, draw a picture of the graph.

   (a) $V(G) = \{a, b, c, d, e\}$ and $E(G) = \{ab, bc, bd, cd, ce, de\}$
   (b) $V(G) = \{a, b, c, d\}$ and $E(G) = \{ab, ac, ad, bc, bd, cd\}$
   (c) $V(G) = \{a, b, c, d, e, f\}$ and
       $E(G) = \{ab, ad, af, bc, be, cd, cf, de, ef\}$

3. Draw two different graphs with 4 vertices and 4 edges.

4. Find the center, radius, and diameter for several paths and prove that the centers of the paths consist either of a single vertex or two adjacent vertices. When does each case occur? Find a way of predicting the radius and diameter of P_n in terms of n.

5. Jordan showed, in 1869, that the center of any tree (not just paths) consists of either a single vertex or two adjacent vertices. He did this by observing that if one deletes all the endvertices of a tree, each of the eccentricities of the remaining vertices decreases by one. Thus the center is unaffected, since the vertices with minimum eccentricity still have minimum eccentricity after this deletion. Upon repeating this deletion process, one must eventually obtain either a single vertex or two adjacent vertices. Do this for the tree in Figure 10-7. Then prove Jordan's claim by validating his reasoning.

6. Find a way to predict the radius and diameter of C_n in terms of n. How can you know in advance that the diameter of C_n is smaller than that of P_n?

7. The *complete graph* K_n consists of n vertices such that each vertex is adjacent to every other vertex in the graph, that is, all pairs of vertices are adjacent. Show that K_n is regular. What is the common degree? Now use a theorem to find the number of edges.

8. Use the following alternate plan to find the number of edges in K_n. Each edge involves selecting a pair of vertices from among n of them. This number should be $C(n, 2)$, that is, the number of ways of selecting two objects from among n distinct objects. We hope you get the same answer as in the previous problem.

9. Let's do the previous problem in a third way. (This is pushing our luck.) Label the vertices 1 through n. Now vertex 1 is adjacent to $n - 1$ vertices since it isn't adjacent to itself. Then it contributes $n - 1$ edges to the total. Now consider vertex 2. It introduces only $n - 2$ new edges since we already counted the edges involving vertex 1. Vertex 3 introduces $n - 3$ new edges, and so on, until vertex $(n - 1)$ introduces 1 new edge. Clearly, vertex n introduces no new edges. Then the total number of edges is the sum of the integers from 1 to $n - 1$, which we have done earlier in this book. Go for it. The moral of this is that there are many ways to solve a math problem. Before we leave these graphs, draw K_5 and show that there is no way to do this without at least two edges crossing one another. Draw K_4 with no edges crossing.

10. Show that the eccentricities of two adjacent vertices are either equal or they differ by one. Draw several graphs and compute eccentricities before you attempt to solve this problem.

11. Show that if we are given vertices x, y, and z in a connected graph G, we have the following inequality: $d(x,y) \leq d(x,z) + d(z,y)$. This is called the *triangle inequality* and says that a shortest journey between two points cannot be made shorter by stopping at a third point. Every cabdriver knows this! Deduce from this inequality that the diameter of a graph cannot be greater than twice the radius.

12. Design a 3-regular graph with 8 vertices. If k is an odd number, why must a k-regular graph have an even number of vertices? (*Hint*: Use Euler's Theorem.)

13. Here is a hard problem. Draw several graphs and label the degree of each vertex. Notice that a degree is repeated in each graph. Show that for any graph with at least two vertices, a degree must occur at least twice, that is, the degrees of the vertices cannot be distinct. [*Hint*: The highest possible degree in a graph consisting of n vertices is $n - 1$. The lowest possible degree is 0. This gives us n distinct numbers. But . . . can a graph with an isolated vertex (a vertex of degree 0) have a vertex of degree $n - 1$?]

14. Use Lagrange's formula to find the height from which an object must be dropped so that its speed when it hits the ground is 200 ft/sec. Repeat the problem using a terminal velocity of 400 ft/sec. Why isn't your second answer twice as big as the first?

15. Write the numbers from 30 to 50 as sums of at most four squares. Can this ever be done in more than one way? Write 100 as the sum of at most four squares in five different ways!

16. List all seven partitions of 5. Now divide them into two groups, the first containing all partitions with 3 or fewer terms, the second containing the partitions where no term is greater than 3. Do the two groups contain any of the same members? Do they contain the same number of partitions?

17. List all the partitions of 8. Now divide them into two groups. In the first group, list all partitions using 4 or fewer terms. In the second group, list those partitions with no terms greater than 4. Verify that both groups have the same number of partitions.

18. Euler called a partition in which all terms are odd, an *odd partition*. He called a partition whose terms were distinct, a *distinct partition*. In the preceding exercise, $3 + 3 + 1 + 1$ is an odd partition but not a distinct one, while $4 + 3 + 1$ is distinct but not odd. (Some partitions, such as $5 + 3$, are both odd and distinct, while others, such as $4 + 4$ are neither.) Euler showed that given any integer, there are the same number of odd partitions as there are distinct partitions. The proof is too difficult to present or to assign as homework here. Verify his assertion for the partitions of 8 in the previous exercise.

Suggestions for Further Reading

1. Dunham, William. *Euler: The Master of Us All.* MAA, Washington, DC, 1999.

2. Euler, Leonhard. "The Königsberg Bridges," *Scientific American*, 189(53), 66-70.

3. Harary, Frank. *Graph Theory.* Addison-Wesley, New York, 1969.

4. Maor, Eli. *E: The Story of a Number.* Princeton University Press, Princeton, NJ, 1998.

5. Roberts, Fred S. "Graph Theory and Its Applications to Problems of Society," CBMS-NSF Monograph No. 29. SIAM, Philadelphia, 1978.

6. Stein, S.K. *Mathematics: The Man Made Universe.* Dover, New York, 2000.

Chapter 11

A Century of Surprises

The nineteenth century saw a critical examination of Euclidean geometry, especially the *parallel postulate* which Euclid took for granted. It says, essentially, that through a given point P not on a given line L, there exists exactly one line parallel to L. Any other line through P will, if extended far enough, meet L. Mathematicians sought a proof of this postulate for 2,000 years even though Euclid presented it as a self-evident idea not requiring proof. They did this because to them it was not self-evident, but instead, they thought, a consequence of previous results and axioms which were self-evident.

After failing to prove the parallel postulate, mathematicians wondered if there was a consistent "alternative" geometry in which the parallel postulate failed. To their amazement, they found two! The secret was to look at curved surfaces. You see, the plane is flat – it has no curvature. (Actually, its curvature is 0.)

Consider the surface of a giant sphere like Earth (approximately). To do geometry, we need a concept analogous to the straight lines of plane geometry. What do straight lines in the plane do? Firstly, the line segment PQ yields the shortest distance between points P and Q. Secondly, a bicyclist traveling from P to Q in a straight line will not have to turn his handlebars to the right or left. His motto will be "straight ahead." Similarly, a motorcyclist driving along the equator between two points will be traveling the shortest distance between them and will appear to be traveling straight ahead, even though the equator is curved. Like his planar counterpoint on the bicycle, our motorcyclist will not have to turn his handlebars to the left or right. The same would hold true if he were to travel along a meridian, which is sometimes called a longitude line. (Longitude lines pass through the North and South Poles.)

Meridians and the equator have in common that they are the intersections of the earth with giant planes passing through the center of Earth. In the case of the equator, the plane is (approximately) horizontal, while for the meridians, the planes are (approximately) vertical. Of course, there are infinitely many other planes passing through the center of Earth which determine many other so-called "great circles" which are neither horizontal nor vertical. Given two points, such as New York City and London, the shortest route is not a latitude line but rather an arc of the great circle formed by intersecting Earth with a plane passing through New York, London, and the center of Earth. This plane is unique since three non-collinear points in space determine a plane, in a manner analogous to the way two points in the plane determine a line.

Geometers call a curve on a surface which yields the shortest distance between any two points on it a *geodesic curve*, or just a *geodesic* for short. This enables us to do geometry on curved surfaces. Imagine a triangle on Earth with one vertex at the North Pole and two others on the equator at a distance $\frac{1}{4}$ of the circumference of the earth. All three angles of this triangle are $90°$, so the angle sum is $270°$!! In fact the angle sum of any spherical triangle is larger than $180°$ and the excess, it was shown, is proportional to its area.

In this geometry, there is no such thing as parallelism. Two great circles must meet in two *antipodal points*, that is, two endpoints of a line passing through the center of the sphere.

An even stranger geometry is needed for surfaces like a *saddle*. Imagine a saddle on a camel placed between its neck and its hump. (See Figure 11-1.) The camel driver is sitting at the bottom of a U-shaped curve determined by the neck and hump. But he is also sitting on an upside down U-shaped curve determined by his legs which wrap partially around the body of the camel. On this kind of surface, parallel geodesics actually diverge! They get farther apart, for example, if they go around different sides of its neck. The stranger part is that through a point P not on a given line L on the surface, there are infinitely many parallel lines. Furthermore, angle sums of triangles on these saddlelike surfaces are less than $180°$.

Figure 11-1

These two geometries prepared mathematicians and physicists for an even more bizarre geometry required by Albert Einstein[1] (1879-1955), whose theory of relativity, in the first half of the twentieth century, would shock the world and alter our conception of the physical universe.

Another interesting development was in the making – the algebra of *vectors*. A vector is best viewed as an arrow. It has magnitude (length) and direction. It is an excellent candidate for representing velocity or force. After all, a speeding car has a numerical speed,

---

[1]According to Einstein, "Teaching should be such that what is offered is perceived as a valuable gift and not as a hard duty."

like 60 mph – but it has a direction, too, like northeast. So we can represent the velocity vector by drawing an arrow of length 60 pointing in the northeast direction, as in Figure 11-2.

Figure 11-2

Letting bold letters such as **u** and **v** represent vectors, mathematicians and physicists wondered how to do algebra with them, that is, how to manipulate them in equations as if they were numbers. The simplest operation is addition, so what is **u** + **v** ? Picture yourself in a sailboat (reading this book), and suppose the wind pushes you due east at 8 mph while the current pushes you due north at 6 mph. The sum of these vectors should reflect your actual velocity, including both magnitude and direction. Since the two velocities (wind and current) act independently, it was realized that the two vectors could be added consecutively, that is, one after the other, as is shown in Figure 11-3, where the tail of the second vector **v** is placed at the head of the first vector **u**. The sum is a vector **w** whose tail is the tail of the first vector and whose head is the head of the second.

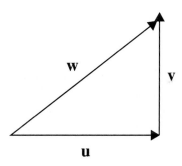

Figure 11-3

The magnitude of **w** is easy to find here since the three vectors form a right triangle. The Pythagorean Theorem tells us that the length of **w** is $\sqrt{6^2 + 8^2} = \sqrt{100} = 10$. The speed of the sailboat is 10 mph. The boat is not traveling exactly northeast because the angle between vectors **u** and **w** is not 45°. The exact angle may be computed using trigonometry. It will be a bit less than 45° since **v** is shorter than **u**.

How do we evaluate sums of three or more vectors? The same way. Place them consecutively so that the tail of each vector coincides with the head of the previous one. The sum will be a vector whose tail is the tail of the first vector and whose head is the head of the last vector. Figure 11-4 demonstrates this for the sum **s** of three vectors.

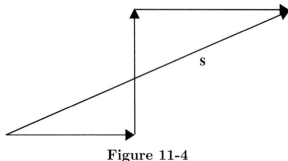

Figure 11-4

Needless to say, all of this applies to forces impinging simultaneously on an object or even an atomic particle. Furthermore, these vectors do not have to lie in the same plane. Fortunately for scientists, the entire analysis can be done in three-dimensional space. Otherwise, vectors wouldn't model the real world.

A vector can be described with the use of *components*, as follows. Place the vector in x, y, z space with its tail at the origin. The coordinates of the head are then taken as the components of the vector. We use the notation $[a, b, c]$ here to distinguish vectors from points, that is, to distinguish components from coordinates. Mathematicians were delighted to discover that the geometric instructions for addition given above simplify greatly to a mere adding of respective components.

Thus if $\mathbf{u} = [a, b, c]$ and $\mathbf{v} = [d, e, f]$, then $\mathbf{u} + \mathbf{v} = [a+d, b+e, c+f]$. This also answered the question, how do we multiply a vector by a number? What would $2 \times \mathbf{u}$ be? It seems that it should correspond to $\mathbf{u} + \mathbf{u}$, or $[a, b, c] + [a, b, c] = [2a, 2b, 2c]$. This suggests that we have the right to distribute a multiplying number to each component of the vector.

Let's summarize these amazing facts.

$$[a, b, c] + [d, e, f] = [a + d, b + e, c + f]$$
$$k \times [a, b, c] = [ka, kb, kc]$$

Here is an example to make this more concrete.

Example 11-1 *Let* $\mathbf{u} = [1, -2, 3]$, $\mathbf{v} = [0, 2, 1]$, *and* $\mathbf{w} = [-2, 1, -2]$. *Find* $\mathbf{u} + \mathbf{v}$, $\mathbf{u} + \mathbf{w}$, $\mathbf{v} + \mathbf{w}$, *and* $2 \times \mathbf{u}$.

Following the facts above, we have

$$\mathbf{u} + \mathbf{v} = [1 + 0, -2 + 2, 3 + 1]$$
$$= [1, 0, 4]$$

and

$$\mathbf{u} + \mathbf{w} = [1 + (-2), -2 + 1, 3 + (-2)]$$
$$= [-1, -1, 1]$$

and

$$\mathbf{v} + \mathbf{w} = [0 + (-2), 2 + 1, 1 + (-2)]$$
$$= [-2, 3, -1]$$

Also,

$$2\mathbf{u} = [2(1), 2(-2), 2(3)]$$
$$= [2, -4, 6]. \quad \blacksquare$$

Having a penchant for generalization, mathematicians of the nineteenth century conjured up an n-dimensional world, called $\mathbf{R}^n$, in which points have n coordinates and vectors have n components! The above laws carry over quite easily to these n-dimensional vectors and yield an interesting theory which most find impossible to visualize. After all, $\mathbf{R}^2$ has two axes which are mutually perpendicular (meet at right angles). $\mathbf{R}^3$ has three axes which are mutually perpendicular. One adds the z-axis to the existing set of axes in the plane to get the three-dimensional scheme of $\mathbf{R}^3$. Now what? How do we add a new axis so that it will be perpendicular to the x-, y-, and z-axis? This is where imagination takes over. We imagine a new dimension that somehow transcends space and heads off into a fictitious world invisible to nonmathematicians.

All of this might have seemed like a game to its founders until Einstein[2] showed that the universe is four dimensional. Time is the fourth dimension and must be taken into account when computing distance, velocity, force, weight, and even length! He posited that large massive objects (like our sun) curve the four-dimensional space around them and cause other objects to follow curved trajectories around them – hence the elliptic trajectory of Earth around the sun. Einstein[3] correctly predicted that light bends in a gravitational field. This was verified during a solar eclipse at a time when Mercury was on the other side of the sun and normally invisible to us. The eclipse, however, rendered it visible and it seemed to be in a slightly different location precisely accounted for by the bending of light in the gravitational field of the sun.

---

[2] "Imagination is more important than knowledge."

[3] "Only two things are infinite, the universe and human stupidity, and I'm not sure about the former."

The nineteenth century was the time in which electromagnetic phenomena puzzled scientists. An electric current in a wire wrapped around a metal rod generated a magnetic field around it. On the other hand, a moving magnet generated a current in a wire. These phenomena were finally described by laws using *vector fields*, spaces in which each point is the tail of a vector whose magnitude and direction vary from point to point. Mathematics again became the language and lifeblood of science.

Vectors are used today to keep track of airplanes flying over the Atlantic, to do computer graphics, to compute stress in architectural structures, and, of course, to describe the forces of the physical world – from the subatomic level all the way to the galactic.

Before we leave the nineteenth century, another puzzle was left unanswered. The speed of light was measured in two directions, one in the direction of the motion of Earth and the other perpendicular to that direction. The shocking fact was that both speeds were the same. It was finally realized that the speed of light seemed independent of the velocity of its source, in contradiction to the findings of Galileo and Isaac Newton. They posited the law of addition of velocities. If a man on a train runs forward at 6 mph and if the train is moving at 60 mph, the speed of the man relative to an observer on the ground is 60 mph + 6 mph, or 66 mph. Why was light exempt from this law? The answer emerged from Einstein's theory that the universe is four dimensional and requires a complicated mathematical scheme of calculation in which the velocity of light, that is, the speed of propagation of electromagnetic energy, denoted c, is constant and is in fact the limiting speed of the universe. Oddly enough, this same constant c plays a role in the conversion of mass into enormous quantities of energy in nuclear reactions, as is predicted by the famous formula $E = mc^2$.

The propagation of light energy was found to have wavelike properties in which energy levels reach high points and low points in a manner somewhat analogous to sound waves in which the density of air varies from very rarified to very dense. It may also be compared to the ripples of water that spread away from a stone thrown into a lake. Water levels rise and fall like blue ocean waves against the seashore. (Is this last sentence a haiku?) Many other phenomena vary in this way such as the height of a weight bobbing up and down on a spring.

The mathematical function that applies to all of these is called the

sine function, and the behavior of the fluctuating quantity is called
sinusoidal. The sine had its origins in the theory of similar triangles
first developed in Ancient Greece. Recall that two triangles are similar if they have the same angles and their sides are proportional. In
the two similar triangles of Figure 11-5, in which the lengths of sides
of the larger triangle are k times the lengths of the corresponding
sides of the smaller, it follows that the two triangles have the same
internal ratios – in other words,

$$\frac{a}{b} = \frac{ka}{kb}$$
$$\frac{a}{c} = \frac{ka}{kc}$$

and

$$\frac{b}{c} = \frac{kb}{kc}$$

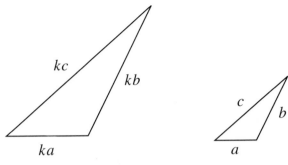

Figure 11-5

Now let's turn our attention to right triangles. Two right triangles are similar if they have the same acute angles (angles less than 90°). Since the two acute angles of a right triangle add up to 90°, all we need to prove similarity is that one of the two angles are the same. For example, if each of two right triangles have a 30° angle, it follows that the other angle must be 60° and the two right angles are similar. Then the ratios of sides are the same in both.

Recall that the sine of $\angle A$ of right triangle $\triangle ABC$ in Figure 11-6, denoted $\sin A$, is the length of the opposite side divided by the length of the hypotenuse, that is,

$$\sin A = \frac{a}{c}$$

The letters used are fairly standard. The vertices are A, B, and C with the right angle at C, while the lowercase letters represent the lengths of the sides directly opposite the vertices. The hypotenuse of the right triangle is unambiguously defined as the longest side of the triangle, or if you wish, the side opposite the right angle. On the other hand, the other two sides are named with reference to the acute angles being considered. From the point of view of angle A, side AC is called the *adjacent* side and BC is called the *opposite* side. From the viewpoint of angle B, these names must be reversed.

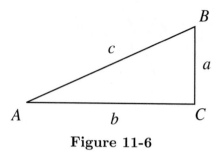

Figure 11-6

Let's remember this. The sine of an acute angle is defined as the ratio of the length of the opposite side to the length of the hypotenuse, that is,

$$\text{sine} = \frac{\text{opposite}}{\text{hypotenuse}}$$

Notice that it is not necessary to specify the particular right triangle containing the given angle. The ratio will be the same in any right triangle since all right triangles containing that angle are similar.

Now imagine a variable right triangle with a hypotenuse c of length 1, as shown in Figure 11-7. As the base angle A grows from $0°$ to $90°$, the length of side a will vary from 0 to 1. Then $\sin A$, which is $\dfrac{a}{c}$, will grow from 0 to 1.

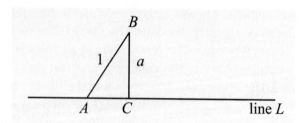

Figure 11-7

Now if we think of A as a stationary point with a long horizontal line through it, two things are obvious.

Firstly,

$$\sin A = \frac{a}{c} = \frac{a}{1} = a$$

or in other words, the sine of angle A is given by a, which simply tells us the height of vertex B above the ground (i.e., above line L).

Secondly, as angle A grows, vertex B describes an arc of a circle of radius one centered at A.

This gave mathematicians a great idea. If angle A is *obtuse* (between $90°$ and $180°$), let its sine still be defined by the height of vertex B – even though we no longer have a right triangle. Continuing with this logic, if angle A is a so-called *reflex angle* (larger than $180°$), vertex B will be under line L and the sine of angle A will be a negative number representing the depth of vertex B. As angle A varies from $0°$ to $360°$, its sine will vary from 0 to 1, back to 0, down to -1, and finally back up to 0. The graph of this newly defined function $y = \sin x$ is shown in Figure 11-8.

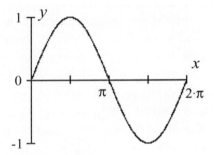

Figure 11-8

This function may be extended past $360°$ in a logical way. If angle A extends to, say, $370°$, it will look exactly like $10°$ and the altitude of vertex B will be the same as it was for $\angle A = 10°$. So the sine is periodic! From $360°$ to $720°$, it simply repeats its basic S-shaped curve. From $720°$ to $1080°$, this same curve will repeat once more, and so on to infinity.

Now if we graph instead, $y = 10 \times \sin x$, we will get almost the same graph. The new graph will have heights which vary between 10 and -10 instead of between 1 and -1. In the function $y = k \times \sin x$, the height k is called the *amplitude* of the sine wave.

On the other hand, how would the graph be affected if the function were $y = \sin(2x)$ or $y = \sin(3x)$? In the first case, as x varies from $0\,°$ to $180\,°$, we would get one complete cycle of the sine wave, since $2x$ would go from $0\,°$ to $360\,°$. Then the complete cycle would occur again as x went from $180\,°$ to $360\,°$, since $2x$ would go from $360\,°$ to $720\,°$. In the case of $y = \sin(3x)$, as x varies from $0\,°$ to $120\,°$, we would get a complete cycle since $3x$ would go from $0\,°$ to $360\,°$. By the time x gets to $360\,°$, we would have three complete cycles. We define the number n in the equation $y = \sin(nx)$ to be the *frequency* of the wave because it tells us how many times the complete cycle occurs as x goes from $0\,°$ to $360\,°$.

Modern-day radios pick up sinusoidal electromagnetic waves in which either

1. the amplitude changes while the frequency stays constant, or

2. the frequency changes while the amplitude stays constant.

The first method of transmission is called *amplitude modulation* (AM), while the second is called *frequency modulation* (FM). The variable amplitude or frequency conveys the precise information needed to re-create the music or speech being broadcast by the radio tower. Who would have believed that trigonometry would help physicists describe the behavior of electromagnetic energy?

The nineteenth century was an age of surprises – technological inventions, mass production, unanswered scientific mysteries, and amazing mathematical concoctions. Little did the world anticipate the events that would unfold in the next century – the age of nuclear energy, space exploration, computers, radio and television, and heart transplants on the one hand and world wars, genocide, totalitarianism, and general worldwide insanity on the other.

EXERCISES

1. Let $\mathbf{u} = [2,2,2]$, $\mathbf{v} = [1,2,0]$, and $\mathbf{w} = [3,2,2]$. Find the following:

   (a) $\mathbf{u} + \mathbf{v} + \mathbf{w}$

   (b) $\mathbf{u} - \mathbf{w}$

   (c) $\mathbf{v} + 2\mathbf{w}$

   (d) $3\mathbf{u} - \mathbf{v}$

   (e) $\mathbf{u} + 2\mathbf{v} - \mathbf{w}$

   (f) $\mathbf{u} - \mathbf{v} + 2\mathbf{w}$

   (g) $5\mathbf{u} + \mathbf{v} + \mathbf{w}$

2. How does the fact that the angles of a triangle add up to $180°$ depend on the validity of the parallel postulate?

3. If two lines, L and M, are parallel to a third line, N, where all three lines are *coplanar* (in the same plane), prove that lines L and M are parallel. You may use the following definition of parallelism. Two lines in the same plane are parallel if they

don't intersect no matter how far they are extended. How would you define parallelism in three-dimensional space? It clearly does not suffice that they do not intersect.

4. Using the North Pole, a segment of the equator of Earth, and segments of two meridians (longitudinal lines), create a triangle whose angle sum is $300°$.

5. Using the geometric approach to adding vectors, explain why $2 \times \mathbf{v} = \mathbf{v} + \mathbf{v}$ is twice as long as $\mathbf{v}$ and points in the same direction, that is, is parallel to $\mathbf{v}$. Can you extend this logic to $3 \times \mathbf{v}$? How about $n \times \mathbf{v}$?

6. Explain geometrically why $\mathbf{v} + (-\mathbf{v}) = \mathbf{0}$. Reminder: The zero vector, $\mathbf{0}$, has neither direction nor magnitude. Conclude that the vector $-\mathbf{v}$ is parallel to $\mathbf{v}$ and has the same magnitude but points in the opposite direction. How would you interpret $-2 \times \mathbf{v}$?

7. If the current in a lake is 8 mph due east and the wind pushes a sailboat 15 mph due north, how fast will the sailboat travel under the influence of the current and the wind? Use a vector diagram and the Pythagorean Theorem.

8. The *cosine* of an acute angle in a right triangle is the length of the adjacent side divided by the length of the hypotenuse. Explain why the cosine of the angle the course of the sailboat makes with the easterly direction in the previous exercise is $\frac{8}{17}$. Using a calculator, the inverse cosine function will then give you that angle.

9. In the right triangle ABC, whose right angle is at the vertex C, let the hypotenuse have length c, and let the sides opposite angles A and B have lengths a and b, respectively. Observe that

$$\sin A = \frac{a}{c}$$

and

$$\cos A = \frac{b}{c}$$

Using this fact and the Pythagorean Theorem, show that

$$(\sin A)^2 + (\cos A)^2 = 1$$

This is true for *any* angle!

10. In the right triangle of the preceding exercise, notice that angles
    A and B are *complementary*, that is, they add up to $90°$. Show
    that the sine of an acute angle (less than $90°$) is the same as
    the cosine of its complement. Deduce a similar statement about
    the cosine of an acute angle.

11. Consider a right triangle whose legs are each unit length. Show
    that the hypotenuse has length $\sqrt{2}$. How do you know that the
    acute angles are each equal to $45°$? Use this to determine the
    sine and cosine of $45°$.

12. Using the previous exercise, find the legs of a $45°$, $45°$, $90°$
    right triangle if its hypotenuse has a length of 1000 ft.

Suggestions for Further Reading

1. Cwiklik, Robert. *Albert Einstein and the Theory of Relativity*, Barrons Solution Series. Barrons Juveniles, New York, 1987.

2. Einstein, Albert. *World As I See It*. Citadel Press, New York, 1993.

3. Einstein, Albert. *Ideas and Opinions*. Bonanza Books, New York, 1986.

4. Krause, E. F. *Taxicab Geometry: An Adventure in Non-Euclidean Geometry*. Dover, New York, 1986.

5. Ryan, Patrick J. *Euclidean and Non-Euclidean Geometry*. Cambridge University Press, Cambridge, MA 1986.

Chapter 12

Ones and Zeros

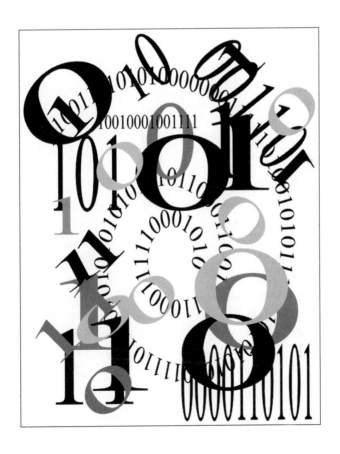

We begin our study of the amazing computer age with a simple idea whose origins date back to Babylon and Egypt. The Babylonians developed the first position system using the rather large base of sixty. Our Hindu-Arabic numbers are based[1] on this idea but employ the simpler base-ten. Thus the number $345,672.568$ means $3 \times \mathbf{100,000} + 4 \times \mathbf{10,000} + 5 \times \mathbf{1000} + 6 \times \mathbf{100} + 7 \times \mathbf{10} + 2 \times \mathbf{1} + 5 \times \frac{1}{\mathbf{10}} + 6 \times \frac{1}{\mathbf{100}} + 8 \times \frac{1}{\mathbf{1000}}$.

Notice that the bold numbers are powers of ten. They get larger by a factor of ten as we move to the left, or, alternatively, get smaller by a factor of ten as we move to the right. This enables us to write any number no matter how large or small using a mere ten digits. It also facilitates the algorithms that we use to add, subtract, multiply, and divide numbers with relative ease. It enables us to carry and borrow in these four operations – techniques we learn in elementary school.

Now recall that Egyptian priests developed a clever multiplication algorithm that involved doubling. We expressed the smaller number as a sum of powers of two, that is, 1, 2, 4, 8, 16, and so on. They did 7×13, for example, by writing 7 as $1 + 2 + 4$. Then we have

$$7 \times 13 = (1 + 2 + 4) \times 13$$
$$= 1 \times 13 + 2 \times 13 + 4 \times 13$$
$$= 13 + 26 + 52$$
$$= 91$$

The Egyptians figured out that any number can be written as a sum of powers of two. We illustrate this for the first 27 numbers in Figure 12-1. Notice that some powers of two are skipped, but powers of two are never multiplied as they are in base-ten where 23 means $2 \times 10 + 3$, that is, the ten is multiplied by two – it is used twice. (23 is $10 + 10 + 3$.) This extraordinary fact makes it desirable to use two as a base of a number system. We call it the *binary system.*

To express any number, we only need ones and zeros! We place them in columns whose place values are powers of two, just as we do in any other base (where the place values are powers of the base). Thus 110010 in the binary system means

$$1 \times 32 + 1 \times 16 + 0 \times 8 + 0 \times 4 + 1 \times 2 + 0 \times 1 = 32 + 16 + 2 = 50$$

Notice that 50 requires only two places, while its binary equivalent 110010 requires six. Binary numbers, however, require only

---

[1]No pun intended.

$$1 = 1$$
$$2 = 2$$
$$3 = 2 + 1$$
$$4 = 4$$
$$5 = 4 + 1$$
$$6 = 4 + 2$$
$$7 = 4 + 2 + 1$$
$$8 = 8$$
$$9 = 8 + 1$$
$$10 = 8 + 2$$
$$11 = 8 + 2 + 1$$
$$12 = 8 + 4$$
$$13 = 8 + 4 + 1$$
$$14 = 8 + 4 + 2$$
$$15 = 8 + 4 + 2 + 1$$
$$16 = 16$$
$$17 = 16 + 1$$
$$18 = 16 + 2$$
$$19 = 16 + 2 + 1$$
$$20 = 16 + 4$$
$$21 = 16 + 4 + 1$$
$$22 = 16 + 4 + 2$$
$$23 = 16 + 4 + 2 + 1$$
$$24 = 16 + 8$$
$$25 = 16 + 8 + 1$$
$$26 = 16 + 8 + 2$$
$$27 = 16 + 8 + 2 + 1$$

Figure 12-1: The First Twenty-seven Numbers Expressed as Sums of Powers of Two

two digits, 0 and 1, while our everyday decimal numbers need ten. The fact that there are only two digits makes binary arithmetic suitable for computers. A circuit has exactly two states. It is either on or off. Computers recognize *on* as 1 and *off* as 0. We will return to this after we examine how various operations are performed in the binary system.

Example 12-1 *Convert 43 to binary.*

To convert a number to binary we divide by two repeatedly until we obtain a quotient of zero. Thus,

$$43 \div 2 = 21 \ R \ 1$$
$$21 \div 2 = 10 \ R \ 1$$
$$10 \div 2 = 5 \ R \ 0$$
$$5 \div 2 = 2 \ R \ 1$$
$$2 \div 2 = 1 \ R \ 0$$
$$1 \div 2 = 0 \ R \ 1$$

Now the solution is the remainders in reverse order, giving us that 43 is 101011 in binary. ∎

Example 12-2 *Convert 110110 to our base-ten system.*

Each place value is a power of 2 starting with 2^0 on the right, so 110110 in binary would be equivalent to

$$1 \times 2^5 + 1 \times 2^4 + 0 \times 2^3 + 1 \times 2^2 + 1 \times 2^1 + 0 \times 2^0$$
$$= 1 \times 32 + 1 \times 16 + 0 \times 8 + 1 \times 4 + 1 \times 2 + 0 \times 1$$
$$= 32 + 16 + 4 + 2$$
$$= 54 \text{ in base-ten} \quad ∎$$

Let's take a look at binary addition. It is clear that $0 + 1 = 1$ and $1 + 0 = 1$, not to mention the fact that $0 + 0 = 0$. On the other hand, $1 + 1 = 10$, since 2 is written 10 in binary. This entails leaving a 0 in the column in which we are working and then carrying a 1 to the next column – exactly what we would do if we were adding a 4 and a 6 in a column of a decimal addition problem. Let's add 110 and 101. The work is shown in Figure 12-2.

```
     110
+    101
    1011
```

Figure 12-2

The first (or *ones*) column is easy, as is the second (or *twos*) column. In either case, $0 + 1 = 1$. The third (or *fours*) column involves the fact that $1 + 1 = 10$. So we put down a 0 in this column and then carry the 1 to the next (or *eights*) column and put it down, yielding the final answer.

The next example is slightly more complicated, as Figure 12-3 shows.

Figure 12-3

Column 1 says that $1 + 1 = 10$, so we put down a 0 and carry a 1 to the next column. Then in column 2, we are adding three ones. The sum, 3, is written 11 in binary so we put down a 1 and carry a 1. The next column adds up to 2 (with the carried 1), or 10 in binary, so we put down a 0 and carry a 1. The next column including the carried 1 adds up to 3 which tells us to put down a 1 and carry a 1 which we put down, and we have our final answer . . . . That was exhausting.

The point is that computers do this using tiny electrical "boxes" that are designed to perform all of this automatically. If a single current comes in, the box is *on*. If two currents come in, the box is *off* but sends a current to the box to its left (metaphorically speaking). This is analogous to $1 + 1 = 10$. If three currents come in, the box is on and sends out a current to its left, analogous to $1 + 1 + 1 = 11$, and so on. In short, addition is now done electronically! And circuits travel almost at the speed of light, which is approximately 186,000 miles per second. As computer circuitry can be measured in mere inches, you may draw the appropriate conclusion about the incredible number crunching speed of the computer.

The computer handles text by converting each letter to a binary number. The same goes for punctuation, the return, shift or tab command, and every other word processing order. An enormous book is stored as a huge succession of ones and zeros!

Since ones and zeros represent pulses and the absence of pulses, digital communication relies, also, on binary arithmetic. Cables carry binary signals around the world representing pictures, music, text, and all other types of media. We have seen only the beginning of the miracles of the computer revolution. The dissemination of information made possible through the Internet, for example, boggles the mind. It will dwarf the changes wrought by the printing press of the 1450s. Stay tuned.

A surprising turn in twentieth-century mathematics was the use of number theory in *cryptography* – the science of coding and decoding information. The National Security Agency employs many mathematicians in the task of creating foolproof codes for military data and breaking enemy codes. There are many stories about the use of cryptography by the United States and Britain in World War II. In fact, cryptography is as old as Ancient Rome. Julius Caesar used a code to transmit state secrets and so did Queen Elizabeth 1600 years later. Coding, or *encryption,* is needed daily today for such mundane matters as paying a bill with a credit card over the Internet or withdrawing cash from an ATM machine. Believe it or not, modular arithmetic started by Fermat plays a key role in the process. So do very large numbers which are products of two primes. The process of factoring a 300-digit number into primes can take many decades even using supercomputers. If your enemy doesn't have your book of primes, you need not even scramble the 300-digit composite number. By the time your enemy factors it, no one will be alive who remembers when, how, or why the war started.

The twentieth century saw the rise of a new field of study called *operations research*. It answers the question: How do I perform a complicated task efficiently? It draws on many traditional fields of mathematics and features practical solutions to seemingly impossible problems. It arose in response to giant tasks such as supplying thousands of wholesale customers with a product, such as gasoline, or shoveling the streets of a large snowbound city with a limited number of snowplows scattered in different locations, or sending bombers from several airfields on simultaneous missions over different targets hundreds of miles apart.

The penalties for not planning a huge task efficiently range from prohibitive costs and utter chaos to complete failure. Imagine catering a party for one hundred guests and not being concerned with the order in which the food must be prepared. This is a childishly simple task when compared with running a business, employing hundreds of workers, which produces and distributes a product to several hundred retail outlets. The law of the free market is, operate efficiently or perish!

The widespread use of computers has enabled planners to perform millions of calculations in a matter of seconds, but the programs must be well thought through. Consider the problem of sorting 20,000,000 names alphabetically. This requires much more than 20,000,000 operations since each name has to be compared with every other name. It turns out that much time is saved by splitting the list into two lists of 10,000,000 names each, sorting them, and then shuffling them into one final list.

The common element of most problems seems to be the necessity of finding a clever algorithm. Suppose you want to determine whether a large number is prime. You might start by seeing if 2 goes into it. If yes, then it's not a prime. If not, try 3. If this goes into it, it isn't prime. If not, try 4, and so on, all the way until you reach the original number, right? Wrong! This will take an extremely long time for a large number. In fact, it would seem as if it took forever.

Here is a great time saving trick. (Never forget the famous equation *time = money*.) Suppose the integer x is a factor of N. Then there must be an integer y such that $x \times y = N$. For example, 5 is a factor of 100. Then, $5 \times 20 = 100$. Notice that $\sqrt{100} = 10$ and one of our factors, 5, is less than 10, while the other factor, 20, is larger than 10.

Let's see if we have a general rule here. 4 is a factor of 100, and $4 \times 25 = 100$. Sure enough, 4 is less than, and 25 is greater than 10. In general, we have the following observation, for which it will be necessary for you to recall that $<$ means *is less than* and $>$ means *is greater than*.

If $N = x \times y$, where $x < y$, then $x < \sqrt{N}$ and $y > \sqrt{N}$.

This makes it unnecessary to check numbers larger than the square root of the number we are testing for primality. If we fail to find a factor by the time we get to the square root, then we won't find a factor larger than the square root.

Let's illustrate this idea. Let's see if 89 is a prime number. We only need to test the numbers up to 9, since $\sqrt{89}$ is "9 point something" (i.e., between 9 and 10). Since the numbers from 2 to 9 do not go into 89 evenly, it is a prime number. If we tested a larger number such as 10,100, we would only need to try possible factors up to 100, since $\sqrt{10,100}$ is approximately 100.5.

We can improve our algorithm by observing that if 2 is not a factor, then we are dealing with an odd number. Hence it has no even factors and we can skip 4, 6, 8, ... saving lots of time and trouble.

The moral of the story is that mathematics is called into play in almost every facet of human endeavor – from distributing cereals to supermarkets to sending a rocket into space.

1. Convert the following decimal numbers to binary form:

    (a) 73

    (b) 29

    (c) 370

    (d) 800

    (e) 31

    (f) 2.5

    (g) 98.75

    (h) 0.125

2. Convert the following binary numbers to decimal form:

    (a) 111

    (b) 10101

    (c) 10000

    (d) 110110

    (e) 11.1

    (f) 11111

    (g) 10.11

    (h) 1.01

    (i) 1001.111

3. Add the following binary numbers:

    (a) $111 + 1010 =$

    (b) $10101 + 11101 =$

    (c) $100 + 10 + 1 =$

    (d) $1111111 + 1 =$

    (e) $0.01 + 0.1 =$

    (f) $0.1 + 0.11 =$

    (g) $100 + 100 =$

    (h) $1000 + 1000 =$

4. Try the following binary subtraction problems. Good luck.

    (a) $1111 - 11 =$

    (b) $1101 - 100 =$

    (c) $110 - 1 =$

    (d) $10000 - 111 =$

    (e) $11011 - 110 =$

    (f) $0.110 - 0.101 =$

5. Attempt the following binary multiplication problems by imitating the algorithm for decimal numbers:

    (a) $111001 \times 10 =$

    (b) $101010 \times 11 =$

    (c) $11100 \times 110 =$

    (d) $10111 \times 101 =$

6. Placing a 0 at the end of a decimal number like 345 has the effect of multiplying it by 10. Why? In a similar manner, what is the effect of placing a 0 at the end of a binary number? What about placing two 0's at the end of a binary number? Generalize to an arbitrary number of 0's. Find a quick test to determine whether a binary number is odd or even.

7. What is the effect of shifting the decimal point of a decimal number several places to the left? (What happens to 347.883 when we change it to 3.47883, for example?) Discuss the analogous situation with binary numbers.

8. In the world of decimals, why does 0.999... (where the "..." means forever) equal 1? What is the value in binary notation of 0.111...?

9. Factor the following numbers (all the way down to their prime factors). Which of the numbers are prime?

   (a) 3000

   (b) 775

   (c) 83

   (d) 226,000

   (e) 6! (recall, $6! = 6 \times 5 \times 4 \times 3 \times 2 \times 1$)

   (f) 6^{10}

   (g) 20×55

   (h) 101

10. In the decimal system, a 1 followed by n 0's has the value 10^n. Illustrate this with a few examples and then explain why it is true. Show that a 1 followed by n 0's in the binary system has the value 2^n. Give several examples.

11. Discuss the similarities and differences between the Babylonian number system and the binary system.

12. *Ternary numbers* like 201120 use three as the base. Why are 0, 1, and 2 the only digits in this system? What is the value of 201120? Determine a procedure for converting decimal numbers into ternary form. Extend the observation of Exercise 10

to ternary numbers like 10000. Evaluate the ternary numbers 0.1 and 0.01.

13. *Octal numbers* use 8 as a base. Find the value of the octal number 31, and use your answer to explain the math joke Dec. 25 = Oct. 31 (after your hysterical laughter subsides, of course). Which digits are utilized by the octal system?

14. If x goes into y and y goes into z, how do you know that x goes into z? (Example: 5 goes into 20, and 20 goes into 80. Why does it follow that 5 goes into 80?) Using this idea, explain why a mathematician testing a number N for primality, after finding that 2 does not go into N, can safely assume that no even number goes into it. If, in addition, 3 does not go into N, why can all multiples of 3 (i.e., 6, 9, 12, ...) be ruled out as possible factors of N? Now see if you can show that we only have to test the prime numbers up to $\sqrt{N}$. If none of them go into N, then it's a prime. Illustrate this by testing 73, using only 2, 3, 5, and 7 as possible factors.

15. For the graph in the accompanying figure, find a cycle containing all the vertices. If the vertices represent towns in a given region which a traveling salesman must visit, he can fly into one of the towns, rent a car, visit each town, return to the first town, and then fly home.

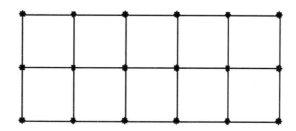

16. Show that if the graph in the previous exercise is modified by deleting the last vertex of each of the three rows, the problem has no solution.

17. Zero is a very interesting number, without which we would have no numbers after nine. Write a short research paper on the history of this fabulous number.

18. *RSA public key* cryptography and the *secure socket layer* are the primary means of securing communications over the World Wide Web. These algorithms use modular arithmetic. Investigate these two algorithms and how they each use modular arithmetic and write your findings in a short paper.

Suggestions for Further Reading

1. Fowler, Mark, and Parekh, Radhi. *Codes and Ciphers* (Usborne Superpuzzles: Advanced Level). E. D. C. Pub., New York, 1995.

2. Gardner, Martin. *Codes, Ciphers and Secret Writing.* Dover, New York, 1984.

3. Ifrah, Georges. *From One to Zero: A Universal History of Numbers.* Viking, New York, 1985.

4. Johnson, Bud. *Break the Code: Cryptography for Beginners.* Dover, New York, 1997.

5. Kaplan, Robert. *The Nothing That Is: A Natural History of Zero.* Oxford Press, New York, 1999.

6. Seife, Charles. *Zero: The Biography of a Dangerous Idea.* Viking, New York, 2000.

Chapter 13

Some More Math before You Go

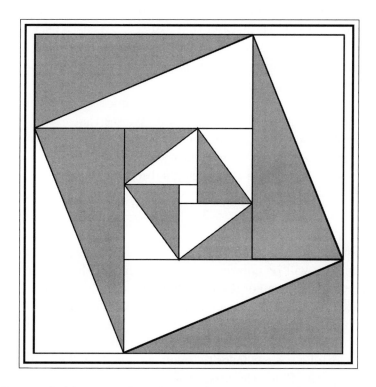

You probably remember the quadratic formula for solving equations of the form

$$ax^2 + bx + c = 0$$

in which a, b, and c represent the so-called constants, or known quantities. We spoke about it earlier. It is easy to determine these constants from a given quadratic. In the equation $2x^2 + 3x - 5$, for example, $a = 2$, $b = 3$, and $c = -5$. At times, b or c can be zero, as in the equations $x^2 - 9 = 0$ and $x^2 - 8x = 0$, respectively. The quadratic formula

$$x = \frac{-b \pm \sqrt{b^2 - 4ac}}{2a}$$

usually yields two different solutions in light of the *plus or minus* sign ($\pm$). The quantity $b^2 - 4ac$ is called the *discriminant*. If it is a negative number, it follows that no real number satisfies the quadratic equation. This is because the square root of a negative number is not real! Mathematicians invented the so-called *imaginary number i* to rectify this situation. $i = \sqrt{-1}$. If one accepts this bizarre invention, every negative number has a square root as we shall now see. Recall, first, that $\sqrt{ab} = \sqrt{a} \times \sqrt{b}$.[1] Then to compute $\sqrt{-9}$, for example, we observe that $-9 = 9 \times (-1)$. Then $\sqrt{-9} = \sqrt{9} \times \sqrt{-1} = 3i$. We won't pursue this idea here but rest assured that imaginary numbers play an important role in both theoretical and applied mathematics today.

Returning to the real world, let's solve a quadratic equation using the formula. Consider $x^2 - 5x + 6 = 0$. We see that $a = 1$, $b = -5$, and $c = 6$. The discriminant is $(-5)^2 - 4 \times 1 \times 6 = 25 - 24 = 1$. Great. It's positive so we will have two solutions, after observing that $\sqrt{1} = 1$. We denote them using subscripts.

$$x_1 = \frac{5 + 1}{2} = 3 \qquad x_2 = \frac{5 - 1}{2} = 2$$

If you plug either of these numbers back into the given quadratic equation, you will be amazed to see that they both work! Now you might wonder how quadratic equations come up in problem solving. Suppose a rectangle has an area of 143 square feet. Moreover, suppose you know that the dimensions differ by 2 feet, that is, the length is two feet longer than the width. To find the dimensions you can let the width be x – our favorite unknown quantity. Then the length must be $x + 2$. Since the area of a rectangle is the product of the length and width, we get the equation

$$x(x + 2) = 143$$

---

[1] This formula is valid as long as at least one of a or b is positive.

which, at first glance doesn't look like a quadratic equation. Let's go to the math lab:

$$x(x + 2) = 143$$

becomes, after distributing the x,

$$x^2 + 2x = 143$$

which, after transposing the 143 to the left side, yields

$$x^2 + 2x - 143 = 0$$

which is as good a quadratic as any we've ever seen! The quadratic formula yields a positive answer $x = 11$, and a negative answer $x = -13$. The second value must be discarded since length cannot be negative (at least not in this world). The positive answer is, of course, correct. The width is 11 and the length, represented by $x + 2$, is 13. The dimensions 11 and 13 yield an area of 143.

Before we present another method of solving quadratics,[2] we must review multiplication of two (simple!) binomial expressions, such as $2x + 1$, or $x - 3$. In fact, let's multiply $2x + 1$ by $x - 3$. The work is arranged in columns just as you would do with ordinary numbers like 34 and 56

$$
\begin{array}{r}
2x + 1 \\
\times \quad x - 3 \\
\hline
-6x - 3 \\
2x^2 + x \\
\hline
2x^2 - 5x - 3 \\
\end{array}
$$

Notice that the first term in the answer, $2x^2$, is the product of the *first* terms of the factors, that is, $2x$ and x, while the *last* term in the answer, -3, is the product of the last terms of the factors, that is, $+1$ and -3. The origin of the middle term, $-5x$, is harder to understand. It is the result of adding the *outer* and *inner* products of the factors when they are written next to each other as $(2x+1)(x-3)$. The outer product is $2x \times (-3)$, or $-6x$, while the inner product is $(+1) \times x$, or $+x$. As the middle column of our chart shows, $-6x$

---

[2]The Babylonians solved quadratic equations by the method known as *completing the square*.

and $+x$ add up to $-5x$. We are assuming that you know how to add signed numbers. Think of consolidating a debt of $-6x$ and an asset of $1x$ (same as x). The result is a debt of only $-5x$, hence the sum of $-5x$. You may remember this by its acronym F.O.I.L. which stands for first, outer, inner, and last.

What is the point of all this? Let's take another look at our earlier quadratic equation $x^2 - 5x + 6 = 0$. Might this be factored into simple expressions of first degree, that is, expressions involving x to the first power? Well x^2 is just x times x. On the other hand, $6 = 6 \times 1 = (-6) \times (-1) = 3 \times 2 = (-3) \times (-2)$. How do we decide between the possible answers

$$(x + 6)(x + 1)$$
$$(x - 6)(x - 1)$$
$$(x + 3)(x + 2)$$
$$(x - 3)(x - 2)$$

without doing all four multiplications? The answer is easy. The middle term, $-5x$, is clearly the result of adding the inner and outer products of the last pair of factors! Aha!

The quadratic may be rewritten as

$$(x - 3)(x - 2) = 0$$

We really have to talk about this. How can the product of two numbers be zero? Easy. One (or both) of them must be zero. Then we can try equating each factor to zero and solving for x. If $x - 3 = 0$, x must be 3, while if $x - 2 = 0$, x must be 2. This agrees perfectly with the solutions obtained earlier using the quadratic formula.

Unfortunately, this method often fails (the method – not the student!). Many quadratics can't be factored using rational numbers (ratios of whole numbers), so the quadratic formula will be around for awhile.

Quadratic expressions often occur in equations that describe a relationship between two variables, such as distance and time in physics, or price and profit in economics. While the graph of the relationship

$$y = ax^2 + bx + c$$

is always a parabola, it is harder to graph than the relatively simple $y = x^2$, treated in an earlier chapter. Just follow these easy steps and you will never fear graphing parabolas ever again. If a is positive, the parabola opens upward. If a is negative, the parabola is upside down (like the trajectory of a football).

Every parabola has an *axis of symmetry* – a vertical line acting like a mirror through which one-half of the parabola seems to be the reflection of the other half. The equation of the axis of symmetry is

$$x = \frac{-b}{2a}$$

Now we look for the *vertex*, that is, the lowest or highest point on the parabola. Well we already have the x-coordinate. The vertex is obviously on the axis of symmetry! To get the y-coordinate, insert this x value into the dreaded quadratic expression.

Finally, pick a few well-chosen x values and find their corresponding y values with the help of the equation and plot them. Then connect the dots and you will have a decent sketch indeed. Let's do an example.

Example 13-1 *Graph the parabola* $y = x^2 - 4x + 1$.

Since $a = 1$ and $b = -4$, the equation of the axis of symmetry is $x = \dfrac{-(-4)}{2\,(1)} = \dfrac{4}{2} = 2$. Plugging this into the parabola's equation, we get

$$y = 2^2 - 4\,(2) + 1 = 4 - 8 + 1 = -3$$

hence, the y value of the vertex is -3. So the vertex is $(2, -3)$. Since a is positive, the parabola opens upward. We recommend a table of values containing numbers on either side of the axis of symmetry. Thus we will use the x values 0, 1, 2, 3, and 4.

x	y
0	1
1	-2
2	-3
3	-2
4	1

To graph the parabola, we draw the axis of symmetry and then plot these points.

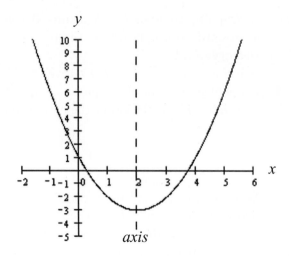

$axis$

We now turn our attention to *simultaneous equations*. Suppose you were told to solve the equation $x + y = 10$. Presumably, *solve* means "find values of x and y which satisfy the equation." Well, there are infinitely many pairs of values which satisfy this equation. Aside from the obvious positive integer pairs like $x = 3$, $y = 7$, we have integer solutions such as $x = -5$, $y = 15$, or $x = -100$, $y = 110$. Then there are decimal solutions like $x = 2.7$, $y = 7.3$, and so on and so forth. You get the idea. Since the graph of $x + y = 10$ is a straight line, there are infinitely many points on it. The coordinates of these points satisfy the equation – indeed there are infinitely many "solutions."

On the other hand, suppose, in addition, we require that $x - y = 6$. Shouldn't this narrow things down somewhat? Let's analyze this situation from two points of view. Geometrically, it should be clear that the lines $x + y = 10$ and $x - y = 6$ intersect in exactly one point. After all, the lines are not parallel since they have different slopes. To see this, convert both lines into the form $y = mx + b$, that is, solve for y, yielding $y = -x + 10$ and $y = x - 6$. These lines have slopes -1 and 1, respectively, so they're not parallel. In fact they are *perpendicular* because they have slopes that are negative reciprocals, but we don't need this fact now.

Algebraically, lets add the two equations:

$$x + y = 10$$
$$x - y = 6$$

yielding

$$2x = 16$$

in which case $x = 8$. Plugging this value of x into the first equation above tells us y must be 2. Notice that these values satisfy both equations. Neat. But what if the two equations are

$$2x + 3y = 12$$
$$5x - 4y = 7$$

and adding the equations fails to eliminate one variable?

As the old cliché goes, we have ways of making the equations talk. Let's multiply the first equation by 5, and the second one by 2, yielding

$$10x + 15y = 60$$
$$10x - 8y = 14$$

and x has the same coefficient in both equations. Let's now subtract the second equation from the first, yielding

$$23y = 46$$

from which we see that $y = 2$. You can then actually put this value into any of the equations and solve for x giving us that $x = 3$.

A word of caution. Not every pair of simultaneous equations has a solution. Consider the system (several equations are called a *system*)

$$x + y = 10$$
$$2x + 2y = 30$$

This system has no solution. If $x + y = 10$, we obtain, after doubling both sides, $2x + 2y = 20$. The second equation is inconsistent with the first so we call the entire system *inconsistent*. After all, how can $2x + 2y$ be both 20 and 30 at the same time? On the other hand, what about the system

$$x + y = 10$$
$$2x + 2y = 20$$

which clearly has infinitely many solutions, since they are just the same straight line. Mathematicians are disappointed here and complain that there is no *unique* solution. Both problems stem from the same root problem. The lines $x + y = c$ and $2x + 2y = d$ have the same slope no matter what values c and d have!! The lines are either parallel or coincident, that is, either they don't meet or they meet in infinitely many points because they are the same line! Very interesting.

These systems can be extended to three or more variables and equations. These kinds of problems are very important in operations research in which many constraints present themselves as equations or inequalities in several variables such as time, labor and equipment costs, transportation fees, and so forth.

Gabriel Cramer[3] (1704-1752), who amazingly completed his doctorate at the young age of 18 and was appointed chair of mathematics at the Académie de Calvin in Geneva at 20, presented a method to solve systems of equations in his book *Introduction à l'analyse des lignes courbes algébraique*. This method is known today as *Cramer's Rule*. Cramer's Rule involves the mathematical concept of *determinants*.

A determinant is a function that operates on a square array of numbers. It is usually represented by vertical lines around the array as is done with the absolute value of a number (assuming of course that you remember the absolute value of x, denoted $|x|$). The 2×2 determinant is defined as

$$\begin{vmatrix} a & b \\ c & d \end{vmatrix} = a \times d - b \times c$$

Let's try this with some numbers.

---

[3]Cramer published articles on a wide range of topics including geometry, philosophy, the aurora borealis, the law, and of course, the history of mathematics. He also worked with many famous mathematicians such as Euler and Johann Bernoulli and was held in such high regard that many of them insisted he alone edit their work.

Example 13-2 *Evaluate the determinant* $\begin{vmatrix} 4 & -2 \\ 3 & 2 \end{vmatrix}$.

Following the formula given above, we have

$$\begin{vmatrix} 4 & -2 \\ 3 & 2 \end{vmatrix} = 4 \times 2 - 3 \times (-2)$$
$$= 8 - (-6)$$
$$= 8 + 6$$
$$= 14 \quad \blacksquare$$

Now, what does this have to do with systems of equations, you ask? Well, we'll tell you. Cramer's Rule says that the solution of a system such as

$$ax + by = c$$
$$dx + ey = f$$

can be found by calculating the three determinants D_x, D_y, and D. These three are defined by

$$D_x = \begin{vmatrix} c & b \\ f & e \end{vmatrix}$$

$$D_y = \begin{vmatrix} a & c \\ d & f \end{vmatrix}$$

and

$$D = \begin{vmatrix} a & b \\ d & e \end{vmatrix}$$

The determinant D is called the *determinant of coefficients*, since it contains the coefficients of the variables in the system. Just in case this looks hard to remember, the x-determinant D_x is simply D with the column containing the coefficients of x replaced by the constants from the system, namely, c and f. Similarly for the y-determinant D_y.

Getting back to the system, the solution is

$$x = \frac{D_x}{D}$$

and
$$y = \frac{D_y}{D}$$
provided of course that D is not 0. (Can you explain why?) Let's look at an example using Cramer's glorious rule.

Example 13-3 *Solve the system below using Cramer's Rule.*

$$x + y = 4$$
$$3x + 4y = 5$$

From the system, we get the three determinants we require.

$$D = \begin{vmatrix} 1 & 1 \\ 3 & 4 \end{vmatrix} = 1 \times 4 - 3 \times 1 = 4 - 3 = 1$$

$$D_x = \begin{vmatrix} 4 & 1 \\ 5 & 4 \end{vmatrix} = 4 \times 4 - 5 \times 1 = 16 - 5 = 11$$

and

$$D_y = \begin{vmatrix} 1 & 4 \\ 3 & 5 \end{vmatrix} = 1 \times 5 - 3 \times 4 = 5 - 12 = -7$$

Since $D \neq 0$, the solution is

$$x = \frac{11}{1} = 11 \quad \text{and} \quad y = \frac{-7}{1} = -7 \quad \blacksquare$$

Before we look at algebraic fractions, let's reexamine numerical ones. One can't add $\frac{3}{5}$ and $\frac{5}{6}$ directly. The first involves 3 slices of a pizza cut into 5 equal slices, while 5/6 represents 5 slices of a pizza pie that has been cut into 6 equal slices. The slices aren't the same size! It's like adding 3 pounds and 5 kilograms.

The brilliant solution is to alter the fractions so they have the same denominators – the so-called *least common denominator* (LCD for short). $\frac{3}{5}$ becomes $\frac{18}{30}$ when we multiply its numerator and denominator by 6, while $\frac{5}{6} = \frac{25}{30}$ when we multiply its numerator and denominator by 5. Now

$$\frac{18}{30} + \frac{25}{30} = \frac{43}{30}$$

Let's do this with algebra. Here's an example.

Example 13-4 *Add the following:*

$$\frac{x+1}{x^2} + \frac{2}{x}$$

The problem is that the denominators are different. The LCD is x^2, so the first fraction needs no adjusting. We must, however, multiply the numerator and denominator of the second fraction by x, making it $\frac{2x}{x^2}$. The problem now becomes

$$\frac{x+1}{x^2} + \frac{2}{x} = \frac{x+1}{x^2} + \frac{2x}{x^2} = \frac{x+1+2x}{x^2} = \frac{3x+1}{x^2} \quad \blacksquare$$

At times one encounters equations with fractions where the easiest approach is to multiply both sides of the equation by the LCD of its fractions. Here's an example:

Example 13-5 *Solve the following equation:*

$$\frac{x}{2} + \frac{2}{x} = 4 - \frac{3}{2}$$

The LCD of the fractions is $2x$, so we multiply both sides of this equation by $2x$, remembering to distribute the $2x$ to both fractions on the left, yielding

$$(2x)\frac{x}{2} + (2x)\frac{2}{x} = (2x)4 - (2x)\frac{3}{2}$$

which, after cancellation, becomes

$$x^2 + 4 = 8x - 3x$$

After simplifying the right side and transposing everything to the left, we have the quadratic equation

$$x^2 - 5x + 4 = 0$$

This can be solved by factoring into $(x-4)(x-1) = 0$. Hence the answers are $x = 4$ and $x = 1$. $\blacksquare$

Let's find the equation of a circle of radius R centered at the origin. If (x, y) is a point on the circle, then its distance from the origin must be R. As the right triangle in the figure below shows, we can apply the Pythagorean Theorem to solve this problem.

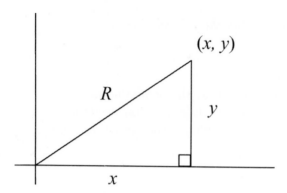

We get the equation $x^2 + y^2 = R^2$. When the radius R is 10, for example, this becomes $x^2 + y^2 = 100$. (Reasoning with circles should not be confused with circular reasoning, by the way.) It seems fitting to end the book with the mysterious circle which so fascinated the thinkers of antiquity. The circle is complete.

Exercises

1. Solve these quadratic equations using the quadratic formula:

    (a) $x^2 - 2x - 20 = 0$

    (b) $2x^2 + 4x - 1 = 0$

    (c) $100x^2 - 9 = 0$

    (d) $8x^2 - 4x = 0$

    (e) $x^2 - x - 1 = 0$

    (f) $x^2 - 4x - 12 = 0$

2. Solve these quadratic equations by factoring. Check your solutions.

    (a) $x^2 - 4x - 12 = 0$

    (b) $x^2 - 2x + 1 = 0$

    (c) $x^2 - 8x + 12 = 0$

    (d) $2x^2 + x - 1 = 0$

    (e) $x^2 + 6x - 40 = 0$

    (f) $x^2 - 10x + 24 = 0$

3. Graph the parabolas of the previous exercise. Be sure to find the axis of symmetry.

4. Graph the following parabolas:

    (a) $y = -x^2$

    (b) $y = 100 - x^2$

    (c) $y = x^2 - 6x$

    (d) $y = 2x^2 + 8x + 10$

    (e) $x = y^2$ (*Hint*: This is a horizontal parabola.)

    (f) $x = y^2 + 6y - 40$

5. Solve the following systems using algebra:

    (a) $x + y = 10$ and $x + 2y = 14$

(b) $2x - y = 7$ and $3x + y = 13$

(c) $5x - 2y = 40$ and $4x + 3y = 55$

(d) $6x - y = 20$ and $x + y = 1$

(e) $\dfrac{x}{2} + \dfrac{y}{3} = 1$ and $x - \dfrac{2y}{3} = 4$

6. Solve the following systems using Cramer's Rule:

   (a) $x + y = 10$ and $x + 2y = 14$

   (b) $2x - y = 7$ and $3x + y = 13$

   (c) $5x - 2y = 40$ and $4x + 3y = 55$

   (d) $6x - y = 20$ and $x + y = 1$

   (e) $\dfrac{x}{2} + \dfrac{y}{3} = 1$ and $x - \dfrac{2y}{3} = 4$

7. Add the following expressions:

   (a) $\dfrac{1}{x^2 - 1} + \dfrac{x}{x + 1} =$

   (b) $\dfrac{10}{x} - \dfrac{x}{x + 3} =$

   (c) $\dfrac{x + 2}{x - 2} + \dfrac{x - 5}{x^2 - 5x + 6} =$

   (d) $\dfrac{1}{x} + \dfrac{x}{x + 1} + \dfrac{3}{2} =$

   (e) $\dfrac{3}{5x + 4} - \dfrac{2}{3x - 1} =$

8. The product of two consecutive numbers is 132. Find the numbers by solving a quadratic equation. (*Hint*: Let the smaller number be x.)

9. Where does the line $x + y = 1$ intersect the parabola $y = x^2$? (*Hint*: Solve the first equation for y in terms of x and plug the resulting expression into the equation of the parabola, yielding a quadratic equation in x. You should get two solutions.)

10. Where does the line $x - y = 1$ intersect the circle $x^2 + y^2 = 1$? (*Hint*: Solve the first equation for y in terms of x and plug the resulting expression into the equation of the circle, yielding a quadratic equation in x. You should get two solutions.)

11. Where does the line $y = x$ intersect the circle $x^2 + y^2 = 1$? (See the hint of the previous exercise.)

12. Where does the parabola $y = x^2$ intersect the circle $x^2 + y^2 = 1$? (*Hint*: Substitute y for x^2 in the equation of the circle, obtaining the quadratic equation $y^2 + y - 1 = 0$. You will get two solutions for y but the negative one must be rejected! This is because $y = x^2$ which is never negative. The positive value for y will yield two x values, for a total of two points.)

13. Write a short paper on two mathematicians. Include information about their lives, their mathematics, and the times in which they lived

Suggestions for Further Reading

1. Frucht, William. *Imaginary Numbers.* John Wiley & Sons, New York, 2000.

2. Nahin, Paul J. *An Imaginary Tale.* Princeton University Press, Princeton, NJ, 1998.

3. Gazale, Midhat J. *Number: From Ahmes to Cantor.* Princeton University Press, Princeton, NJ, 2000.

4. Davis, Martin. *The Universal Computer: The Road from Leibniz to Turing.* W. W. Norton & Co., New York, 2000.

5. Zebrowski, Ernest. *A History of the Circle: Mathematical Reasoning and the Physical Universe.* Rutgers University Press, Piscataway, NJ, 1999.

6. Lay, David C. *Linear Algebra and Its Applications.* Addison-Wesley, New York, 1999.

Appendix A

A Use of Egyptian Fractions

Consider the problem of dividing 4 loaves of bread among 5 people. One solution would be to take $\frac{1}{5}$ off each of the 4 loaves. Then 4 people would each get $\frac{4}{5}$ of a loaf and the fifth person would get 4 cut pieces, each $\frac{1}{5}$ of a loaf. This would be $\frac{4}{5}$ as well. Even though each person gets the same amount, the manner of distribution doesn't seem fair. Four people get a whole piece, while the last gets a lot of little pieces. (See Figure A-1.)

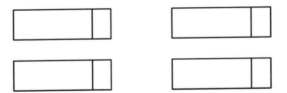

Figure A-1: An Unfair Distribution

Changing $\frac{4}{5}$ to unit fractions, we get

$$\frac{4}{5} = \frac{8}{10} = \frac{5}{10} + \frac{2}{10} + \frac{1}{10} = \frac{1}{2} + \frac{1}{5} + \frac{1}{10}$$

Now, to be fair each person gets half a loaf, one-fifth of a loaf, and one-tenth of a loaf. In order to do this division, cut 2 loaves in halves (that's four of the five needed halves – we still need another half for the fifth person), cut 1 loaf into fifths (that's the five needed fifths), and cut the remaining loaf into tenths but don't cut up the last five-tenths because this is the last needed half. This is truly a fair distribution since every person receives the same number and the same size pieces. (See Figure A-2.) Truly amazing!

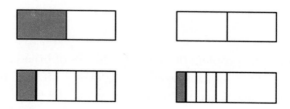

Figure A-2: An Equitable Distribution

Appendix B

Additional Problems

from Babylonia, China, Egypt, Greece, and India

Note: The cubit was a linear measurement from one's elbow to the tip of the longest (middle) finger, usually 17 to 21 inches. The drachma is an ancient Greek coin. Drachmas are still the form of currency today in Greece.

B.1 Pythagorean Problems

1. In a pond, the flower of a water lily is 2 cubits above the water. When it is bent by a gentle breeze, it touches the water at a distance of 4 cubits. Tell the depth of the water. (China) [The depth of the water is 3 cubits.]

2. A chain suspended from an upright post has a length of 9 cubits lying on the ground. When stretched out to its full length so as to just touch the ground, the end is found to be 21 cubits from the post. What is the length of the chain? (China) [The length of the chain is 29 cubits.]

3. A bamboo 36 cubits tall is broken (bent) by the wind so that the top touches the ground 12 cubits from the stem. Tell the height of the break. (Babylonia and China) [The height of the break is 16 cubits.]

4. A snake's hole is at the foot of a pillar which is 24 cubits high with a peacock perched on its summit. Seeing the snake at

a distance of 48 cubits gliding toward its hole, the peacock pounces on it. Say quickly now at how many cubits from the snake's hole they meet, both proceeding an equal distance. (India) [They meet 18 cubits from the hole.]

5. Two magicians live on a cliff of height 40 cubits. There is a stream at a distance of 120 cubits from the foot of the cliff. One magician climbs down and walks to the stream. The other levitates directly up a short distance and then directly to the stream. If both magicians travel the same distance, tell how high the second one flies. (India) [The magician flies 24 cubits high.]

B.2   Rectangle and Square Area Problems

1. The area of a square plus 4 times its side results in 60. Find the length of the side. (Babylonia) [The side is 6.]

2. There exists a rectangle whose area is 180 and whose length is 3 more than its width. Find its dimensions. (Heron of Alexandria) [The rectangle is 12 by 15.]

3. There exists a rectangle with a perimeter of 36 and an area of 80. Find its dimensions. (Heron of Alexandria) [The dimensions are 8 by 10.]

4. The side of a larger square is twice the side of a smaller square plus 5. The sum of the areas is 725. Find the sides of both squares. (Heron of Alexandria) [The sides are 10 and 25.]

5. There exists a rectangle where the length plus width is 21. The area plus the excess of the length over the width is 111. Find the dimensions. (Heron of Alexandria) [The dimensions are 10 by 11 or 9 by 12.]

B.3   Simultaneous Equations

1. There are a number of rabbits and chickens confined in a cage, in all 32 heads and 98 feet. Tell the number of each. (China) [There are 17 rabbits and 15 chickens.]

2. There are a number of rabbits and chickens confined in a cage, in all 30 heads and 98 feet. Tell the number of each. (China) [There are 19 rabbits and 11 chickens.]

3. There are a number of 8-legged spiders and 6-legged ants confined in a cage, in all 35 heads and 244 feet. Tell the number of each. (China) [There are 17 spiders and 18 ants.]

B.4   Diophantine Equations

Diophantus was one of the many ancient mathematicians and philosophers who lived in Alexandria, Egypt. Diophantus posed problems in which there were more unknowns than equations. The solutions are whole numbers.

1. Horses cost 8 coins and cows cost 6 coins. How many of each animal can be purchased for 106 coins? [There are four solutions. If x = the number of horses and y = the number of cows, the solutions are $x = 2$, $y = 15$; $x = 5$, $y = 11$; $x = 8$, $y = 7$; and $x = 11$, $y = 3$.]

2. Ducks can be purchased for 5 drachmas and chickens can be purchased for 4 drachmas. How many of each animal can be purchased for 27 drachmas? [There is only one solution: 3 ducks and 3 chickens.]

3. What combination of quarters and dimes will amount to 78 cents? [There are no solutions.]

4. Horses cost 8 coins, cows cost 5 coins, and 2 pigs for 1 coin. 264 animals cost 264 coins. How many of each animal can be purchased? [There are 6 solutions, the first of which is 2 horses, 26 cows, and 236 pigs.]

5. Ducks can be purchased for 7 drachmas. Chickens can be purchased for 5 drachmas and 3 starlings can be purchased for one drachma. 60 animals cost 60 drachmas. How many of each fowl can be purchased? [There are no solutions.]

6. What combination of half dollars, quarters, and dimes will amount to $6.00 if there are 18 coins? [There is one solution: 9 half dollars, 4 quarters, and 5 dimes.]

B.5 Similar Triangles

1. There is a mountain to be measured. Two rods, each 12 cubits high, are placed in the ground 250 cubits apart. When a man walks back 85 cubits from the rod nearest to the mountain and places his head on the ground, the top of the mountain is just visible through a hole in the top of the rod. The same is true when the man walks back 89 cubits from the rod farthest from the mountain. Find the height of the mountain. [The mountain is 762 cubits high.]

Appendix C

Answers to Selected Exercises

Chapter 1

1. Simple grouping makes it easier to count. The Egyptian number system is an example of a simple grouping system.

5. (a) $73 = 64 + 8 + 1$ (b) $52 = 32 + 16 + 4$
   (c) $98 = 64 + 32 + 2$ (d) $151 = 128 + 16 + 4 + 2 + 1$

7. (a)

*	1	25	
*	2	50	
*	4	100	
	8	200	
*	16	400	
		575	

(b)

	1	82
*	2	164
	4	328
	8	656
	16	1312
*	32	2624
		2788

(c)

*	1	36
*	2	72
*	4	144
*	8	288
*	16	576
		1116

(d)

*	1	107
*	2	214
*	4	428
	8	856
*	16	1712
*	32	3424
		5885

9. (a) $\frac{3}{10} = \frac{1}{5} + \frac{1}{10}$ (b) $\frac{4}{5} = \frac{1}{2} + \frac{1}{5} + \frac{1}{10}$
   (c) $\frac{111}{200} = \frac{1}{2} + \frac{1}{20} + \frac{1}{200}$ (d) $\frac{7}{50} = \frac{1}{8} + \frac{1}{100} + \frac{1}{200}$

10. (a) $\frac{13}{40} = \frac{1}{4} + \frac{1}{20} + \frac{1}{40}$ (b) $\frac{9}{46} = \frac{1}{6} + \frac{1}{138}$

    (c) $\frac{4}{21} = \frac{1}{7} + \frac{1}{21}$ (d) $\frac{4}{11} = \frac{1}{3} + \frac{1}{33}$

11. (a) $\frac{3}{10} = \frac{1}{5} + \frac{1}{10}$ (b) $\frac{4}{5} = \frac{1}{2} + \frac{1}{5} + \frac{1}{10}$

    (c) $\frac{111}{200} = \frac{1}{2} + \frac{1}{20} + \frac{1}{200}$ (d) $\frac{7}{50} = \frac{1}{8} + \frac{1}{100} + \frac{1}{200}$

12. $\frac{1}{n+1} + \frac{1}{n(n+1)} = \frac{n}{n(n+1)} + \frac{1}{n(n+1)} = \frac{n+1}{n(n+1)} = \frac{1}{n}$

13. (a) $\frac{1}{5} = \frac{1}{6} + \frac{1}{30}$ (b) $\frac{1}{11} = \frac{1}{12} + \frac{1}{132}$

    (c) $\frac{1}{10} = \frac{1}{11} + \frac{1}{110}$ (d) $\frac{1}{101} = \frac{1}{102} + \frac{1}{10,302}$.

15. (a) They are both simple grouping systems, but Egyptian hiero-glyphics uses symbols for the powers of ten and the "alienese" uses symbols for the powers of five., that is, $5^0 = 1$, $5^1 = 5$, $5^2 = 25$, $5^3 = 125$, $5^4 = 625$.

    (b) 18 = ###%%, 46 = &####%, 283 = $$&#%%%, 1005 = @$$$#, 3000 = @@@@$$$$

    (c) 3124

16. (a) $75 \div 15 = 5$

1	15	*	75		4
2	30		− 60		+ 1
4	60	*	15		5
			− 15		
			0		

    (b) $156 \div 13 = 12$

1	13		156		8
2	26		− 104		+ 4
4	52	*	52		12
8	104	*	− 52		
			0		

    (c) $80 \div 8 = 10$

1	8		80		8
2	16	*	− 64		+ 2
4	32		16		10
8	64	*	− 16		
			0		

(d) $91 \div 7 = 13$

1	7	*		91			8
2	14		$-$	56			4
4	28	*		35	$+$		1
8	56	*	$-$	28			13
				7			
			$-$	7			
				0			

17. (a) $6,598 \div 37 = 178R12$

1	37			6598			128
2	74	*	$-$	4736			32
4	148			1862			16
8	296		$-$	1184	$+$		2
16	592	*		678			178
32	1184	*	$-$	592			
64	2368			86			
128	4736	*	$-$	74			
				12			

(b) $805 \div 35 = 23$

1	35	*		805			16
2	70	*	$-$	560			4
4	140	*		245			2
8	280		$-$	140	$+$		1
16	560	*		105			23
			$-$	70			
				35			
			$-$	35			
				0			

(c) $528 \div 22 = 24$

1	22			528			16
2	44		$-$	352	$+$		8
4	88			176			24
8	176	*	$-$	176			
16	352	*		0			

(d) $1134 \div 42 = 27$

1	42	*		1134		16
2	84	*	−	672		8
4	168			462		2
8	336	*	−	336	+	1
16	672	*		126		27
			−	84		
				42		
			−	42		
				0		

19. The problem can be written as $x + \frac{x}{7} = 24$ where x is the number. Guessing $x = 7$ yields 8. The "magic number" is $24 \div 8 = 3$. Multiplying this by the original guess gives the correct answer, $x = 21$.

20. (a) $x = 20$ (guess 4) (b) $x = 3.2$ (guess 8) (c) $x = 80$ (guess 16)

Chapter 2

1. Babylonian cuneiform:

(a) ▷ ▽▽▽ / ▷ ▽▽▽   (b) ▷▷ ▽▽▽ / ▷▷ ▽▽▽   (c) ▷ / ▷ ▽▽ / ▷

(d) ▷ ▽▽▽ / ▽   (e) ▷ ▽   (f) ▷▷ ▽▽▽ / ▷ ▽▽

2. (a) $90 = (1, 30)_{60}$ (b) $75 = (1, 15)_{60}$ (c) $3660 = (1, 1, 0)_{60}$
   (d) $7200 = (2, 0, 0)_{60}$ (e) $7325 = (2, 2, 5)_{60}$
   (f) $\frac{3}{4} = \frac{3 \times 15}{4 \times 15} = \frac{45}{60} = (0; 45)_{60}$

3. (a) $47 = (47)_{60}$ (b) $78 = (1, 18)_{60}$ (c) $3662 = (1, 1, 2)_{60}$
   (d) $\frac{2}{3} = \frac{2 \times 20}{3 \times 20} = \frac{40}{60} = (0; 40)_{60}$ (e) $\frac{11}{4} = 2\frac{3}{4} = 2\frac{45}{60} = (2; 45)_{60}$
   (f) $\frac{21}{15} = 1\frac{6}{15} = 1\frac{24}{60} = (1; 24)_{60}$

4. (a) $(1, 1)_{60} = 1 \times 60^1 + 1 \times 60^0 = 1 \times 60 + 1 \times 1 = 60 + 1 = 61$
   (b) $(1, 1, 0; 10)_{60} = 1 \times 60^2 + 1 \times 60^1 + 0 \times 60^0 + 10 \times \frac{1}{60^1} =$
   $1 \times 3600 + 1 \times 60 + 0 \times 1 + 10 \times \frac{1}{60} = 3600 + 60 + 0 + \frac{1}{6} = 3660\frac{1}{6}$
   (c) $(1; 30)_{60} = 1 \times 60^0 + 30 \times \frac{1}{60^1} = 1 \times 1 + 30 \times \frac{1}{60} = 1 + \frac{1}{2} = 1\frac{1}{2}$
   (d) $(1, 2, 3; 30, 20)_{60} = 1 \times 60^2 + 2 \times 60^1 + 3 \times 60^0 + 30 \times \frac{1}{60^1} +$

$20 \times \frac{1}{60^2} = 1 \times 3600 + 2 \times 60 + 3 \times 1 + 30 \times \frac{1}{60} + 20 \times \frac{1}{3600} =$
$3600 + 120 + 3 + \frac{1}{2} + \frac{1}{180} = 3723\frac{91}{180}$ (e) $(1, 0; 0, 1)_{60} = 1 \times 60^1 +$
$0 \times 60^0 + 0 \times \frac{1}{60^1} + 1 \times \frac{1}{60^2} = 1 \times 60 + 0 \times 1 + 0 \times \frac{1}{60} + 1 \times \frac{1}{3600} =$
$60 + \frac{1}{3600} = 60\frac{1}{3600}$

5. Convert to base-10. (a) $(1, 3)_{60} = 1 \times 60^1 + 3 \times 60^0 = 1 \times 60 +$
   $3 \times 1 = 60 + 3 = 63$ (b) $(3, 1, 2; 10)_{60} = 3 \times 60^2 + 1 \times 60^1 +$
   $2 \times 60^0 + 10 \times \frac{1}{60^1} = 3 \times 3600 + 1 \times 60 + 2 \times 1 + 10 \times \frac{1}{60} =$
   $10800 + 60 + 2 + \frac{1}{6} = 10,862\frac{1}{6}$ (c) $(1, 1, 1; 6, 36)_{60} = 1 \times 60^2 +$
   $1 \times 60^1 + 1 \times 60^0 + 6 \times \frac{1}{60^1} + 36 \times \frac{1}{60^2} = 1 \times 3600 + 1 \times 60 +$
   $1 \times 1 + 6 \times \frac{1}{60} + 36 \times \frac{1}{3600} = 3600 + 60 + 1 + \frac{1}{10} + \frac{1}{100} = 3661\frac{11}{100}$
   (d) $(4, 3; 0, 0, 1)_{60} = 4 \times 60^1 + 3 \times 60^0 + 0 \times \frac{1}{60^1} + 0 \times \frac{1}{60^2} + 1 \times \frac{1}{60^3} =$
   $4 \times 60 + 3 \times 1 + 0 \times \frac{1}{60} + 0 \times \frac{1}{3600} + 1 \times \frac{1}{216,000} = 240 + 3 +$
   $\frac{1}{216,000} = 243\frac{1}{216,000}$ (e) $(0; 30, 23)_{60} = 30 \times \frac{1}{60^1} + 23 \times \frac{1}{60^2} =$
   $30 \times \frac{1}{60} + 23 \times \frac{1}{3600} = \frac{1}{2} + \frac{23}{3600} = \frac{1823}{3600}$

6. Find the hypotenuse of the right triangle whose legs are given.
   (a) $c = \sqrt{7^2 + 24^2} = \sqrt{625} = 25$
   (b) $c = \sqrt{9^2 + 40^2} = \sqrt{1681} = 41$

7. Rules for multiplying and dividing by 60 in base-60 are analogous to multiplying and dividing by 10 in base-10, that is, (a)
   when multiplying by 60 in base-60 move the sexagesimal point
   (;) one place to the right. (b) when dividing by 60 in base-60
   move the sexagesimal point (;) one place to the left.

8. (a) The averaging method: Start with a guess of 14 which
   is too small and divide, $200 \div 14 = 14.2857142857\ldots$ which
   is too big. Average the two $(14 + 14.2857142857\ldots)/2 =$
   $14.1428571428\ldots$ This is closer to the correct answer than before. Doing it again, $200 \div 14.142857142857\ldots =$
   $14.1414141414\ldots$. Averaging these last two gives
   $14.1421356421\ldots$. This is extremely close to the correct answer of $14.142135637\ldots$. (b) Using the formula, $S = 14$
   and $E = 4$. Substituting gives $\sqrt{200} = 14 + \frac{4}{2\times14} - \frac{4^2}{8\times14^3} =$
   $14.1421282799\ldots$.

9. All the fractions can be rewritten using a denominator of 60
   which means they will have one "sexagesimal place."
   (a) $\frac{1}{5} = \frac{12}{60} = (0; 12)_{60}$ (b) $\frac{1}{6} = \frac{10}{60} = (0; 10)_{60}$

(c) $\frac{1}{10} = \frac{6}{60} = (0;6)_{60}$ (d) $\frac{1}{12} = \frac{5}{60} = (0;5)_{60}$

(e) $\frac{1}{15} = \frac{4}{60} = (0;4)_{60}$ (f) $\frac{1}{20} = \frac{3}{60} = (0;3)_{60}$

(g) $\frac{1}{30} = \frac{2}{60} = (0;2)_{60}$ (h) $\frac{1}{60} = (0;1)_{60}$

10. Division in base-60.

(a) $13 \div 5 = 13 \times \frac{1}{5} = 13 \times (0;12)_{60} = (2;36)_{60}$ (b) $19 \div 10 = 19 \times \frac{1}{10} = 19 \times (0;6)_{60} = (1;54)_{60}$ (c) $7 \div 24 = 7 \times \frac{1}{24} = 7 \times (0;2,30)_{60} = (0;17,30)_{60}$ (d) $23 \div 12 = 23 \times \frac{1}{12} = 23 \times (0;5)_{60} = (1;55)_{60}$

11. Division in base-60.

(a) $29 \div 16 = 29 \times \frac{1}{16} = 29 \times (0;3,45)_{60} = (1;48,45)_{60}$ (b) $80 \div 15 = 80 \times \frac{1}{15} = 80 \times (0;4)_{60} = (5;20)_{60}$ (c) $5 \div 24 = 5 \times \frac{1}{24} = 5 \times (0;2,30)_{60} = (0;12,30)_{60}$ (d) $17 \div 12 = 17 \times \frac{1}{12} = 17 \times (0;5)_{60} = (1;25)_{60}$

12. Square roots by Babylonian averaging method.

(a) Start with guess 3 (which is too big). After two iterations, we get $\sqrt{8} \cong 2.828431372\ldots$. (b) Start with guess 4 (which is too big). After two iterations, we get $\sqrt{15} \cong 3.872983870\ldots$. (c) Start with guess 6 (which is too small). After two iterations, we get $\sqrt{37} \cong 6.082762557\ldots$. (d) Start with guess 12 (which is too small). After two iterations, we get $\sqrt{145} \cong 12.04159457\ldots$.

13. Square roots by Babylonian formula: (a) $S = 2$ and $E = 4$. Substituting gives $\sqrt{8} \cong 2 + \frac{4}{2\times 2} - \frac{4^2}{8\times 2^3} = 2.75$. (b) $S = 3$ and $E = 6$. Substituting gives $\sqrt{15} \cong 3 + \frac{6}{2\times 3} - \frac{6^2}{8\times 3^3} = 3.83333333\ldots$. (c) $S = 6$ and $E = 1$. Substituting gives $\sqrt{37} \cong 6 + \frac{1}{2\times 6} - \frac{1^2}{8\times 6^3} = 6.0827546296\ldots$. (d) $S = 12$ and $E = 1$. Substituting gives $\sqrt{145} \cong 12 + \frac{1}{2\times 12} - \frac{1^2}{8\times 12^3} = 12.041594328\ldots$.

14. Pythagorean Triples:

m	(a,b,c)
2	(3, 4, 5)
3	(8, 15, 17)
4	(5, 12, 13)
5	(12, 35, 37)
6	(7, 24, 25)

15. Pythagorean Triples

m	(a, b, c)
7	(16, 63, 65)
8	(9, 40, 41)
9	(20, 99, 101)
10	(11, 60, 61)

Chapter 3

1. (i) Since AC and DB are parallel and AB traverses these parallel lines, $\angle A = \angle B$ by alternate interior angles. (ii) Since the vertical angles at E are equal and $\angle C = \angle D = 90°$, the two triangles $\triangle EAC$ and $\triangle EDB$ have two angles equal. Therefore, the third angle of both must also be equal. Hence, $\angle A = \angle B$.

2. Let $x =$ the smallest angle. Then $x+2x+3x = 180°$ or $x = 30°$.

3. Divide the quadrilateral into two triangles by drawing a diagonal. The sum of the angles of the quadrilateral is the sum of the angles of the two triangles. Thus, $2 \times 180° = 360°$.

4. The smaller angle is $18°$.

5. $540°$

6. Pick a vertex and draw diagonals to all but the two adjacent vertices. In doing so, you make $n-2$ triangles. Yes, the formula holds for $n = 3$ and $n = 4$.

7. The proper factors of 220 are 1, 2, 4, 5, 10, 11, 20, 22, 44, 55, and 110. The proper factors of 284 are 1, 2, 4, 71, and 142.

8. A number is even if it is two times something. Thus, if $n = 2k$ is even, then $n^2 = (2k)(2k) = 4k^2 = 2 \times (2k^2) = 2 \times M$. Hence, n^2 is even.

9. The Pythagorean triples are

   (a)   7, 24, 25

   (b)   20, 21, 29

   (c)   8, 15, 17

   (d)   33, 56, 65

(e)   28, 45, 53

(f)   9, 40, 41

10. Answers will vary.

11. Assume $\sqrt{3} = \frac{a}{b}$ where $\frac{a}{b}$ is reduced. Squaring both sides gives $3 = \frac{a^2}{b^2}$ or more simply $a^2 = 3b^2$. Therefore, a^2 is a multiple of 3 which means a is a multiple of 3. Thus, $a = 3k$ and $a^2 = 9k^2$. Substituting yields $9k^2 = 3b^2$ or $b^2 = 3k^2$. Therefore b^2 is a multiple of 3 and hence b is a multiple of 3. Both a and b are multiples of 3, but $\frac{a}{b}$ is reduced. This is a contradiction.

12. Numerology.

(a) Euclid $= 5 + 3 + 3 + 3 + 9 + 4 = 27 = 9$ (high achiever)

(b) Hypatia $= 8 + 7 + 7 + 1 + 2 + 9 + 1 = 35 = 8$ (strong willed)

(c) Pythagoras $= 7 + 7 + 2 + 8 + 1 + 7 + 6 + 9 + 1 + 1 = 49 = 13 = 4$ (honest)

(d) Euler $= 5 + 3 + 3 + 5 + 9 = 25 = 7$ (naturally talented)

Chapter 4

1. Prime or composite: (a) prime (b) prime (c) composite, 2×165

   (d) composite, 3×1521 (e) composite, $2 \times 22,837$ (f) composite, $2 \times 38,618$ (g) prime (h) prime

2. Any number can be written as $10x + d$, where d is the ones digit. Since $10x$ is divisible by 2, any number is divisible by 2 if its ones digit d is divisible by 2. The only single digits divisible by 2 are 0, 2, 4, 6, and 8. This is useful in determining primes, since an even number greater than 2 cannot be prime.

3. Prime factorizations: (a) $56 = 2^3 \times 7^1$ (b) $24 = 2^3 \times 3^1$ (c) $360 = 2^3 \times 3^2 \times 5^1$ (d) $450 = 2^1 \times 3^2 5^2$ (e) $3200 = 2^7 \times 5^2$ (f) $1000 = 2^3 \times 5^3$

4. See answer to Exercise 2.

5.   (a) $GCD(56, 72) = 8$ (b) $GCD(24, 28) = 4$
     (c) $GCD(100, 360) = 20$ (d) $GCD(25, 450) = 25$
     (e) $GCD(150, 270) = 30$

6. (a) $LCM(24, 35) = 840$ (b) $LCM(24, 28) = 168$
   (c) $LCM(20, 45) = 180$ (d) $LCM(56, 72) = 504$
   (e) $LCM(10, 24) = 120$ (f) $LCM(25, 450) = 450$

7. Every number can be written either as $4n$, $4n + 1$, $4n + 2$, or
   $4n + 3$. Since both $4n$ and $4n + 2$ are even (divisible by 2), they
   cannot represent primes. Thus, a prime can be either $4n + 1$
   or $4n + 3$, as demonstrated by 13 and 19. The first few primes
   of the form $4n + 1$ are 5, 13, 17, 29, 37, 41, 53, 61, 73, 89, and
   97. The first few primes of the form $4n + 3$ are 7, 11, 19, 23,
   31, 43, 47, 59, 67, 71, 79, and 83.

8. (a) $GCD(20, 30) = 10$ (b) $GCD(5, 85) = 5$
   (c) $GCD(18, 45) = 9$ (d) $GCD(40, 60) = 20$
   (e) $GCD(17, 19) = 1$ (f) $GCD(1, 1000) = 1$
   (g) $GCD(22, 33) = 11$ (h) $GCD(24, 48) = 24$
   (i) $GCD(6, 60) = 6$

9. (a) $105 \div 15 = 7 \ R \ 0$. Thus, $GCD(15, 105) = 15$. (b) $36 \div 3 = 12$
   $R \ 0$. Thus, $GCD(3, 36) = 3$.   (c) $400 \div 90 = 4 \ R \ 40$ and
   $90 \div 40 = 2 \ R \ 10$ and $40 \div 10 = 4 \ R \ 0$. Thus, $GCD(90, 400) =$
   10.   (d) $1100 \div 400 = 2 \ R \ 300$ and $400 \div 300 = 1 \ R \ 100$
   and $300 \div 100 = 3 \ R \ 0$. Thus, $GCD(400, 1100) = 100$.   (e)
   $775 \div 335 = 2 \ R \ 105$, $335 \div 105 = 3 \ R \ 20$, $105 \div 20 = 5 \ R \ 5$
   and $20 \div 5 = 4 \ R \ 0$. Thus, $GCD(335, 775) = 5$.

10. Essentially, a triangle is uniquely determined by any one of
    the three postulates ASA, SSS, SAS. For further information,
    consult a reference, such as *Geometry Civilized* by Heilbron
    (pp. 54-57), *Basic Geometry* by G. D. Birkhoff and R. Beatley,
    or the many World Wide Web sites, for example,

    - http://forum.swarthmore.edu/dr.math/problems
    - http://www.math.csusb.edu/courses/m129/tri_congr.html
      or
    - http://userzweb.lightspeed.net/jpaulk/mathhelp.htm

11. Since adjacent angles are supplementary, (see Figure 4-1) $\angle A +$
    $\angle B = \angle B + \angle C$ (= 180). Subtracting $\angle B$ from both sides,
    gives $\angle A = \angle C$.

12. Since M is the midpoint of the line segments $\overline{AB}$ and $\overline{CD}$, it follows that $\overline{AM} \cong \overline{MB}$ and $\overline{CM} \cong \overline{MD}$. Also, the angles at M are vertical angles. Thus, the two triangles $\triangle DMB$ and $\triangle CMA$ are congruent by SAS(side, angle, side). Since they are congruent, corresponding angles are equal.

13. Pythagoras says that $a^2 + b^2 = c^2$ or $b = \sqrt{c^2 - a^2}$. Therefore, the other leg of the two triangles must also be equal. Since all three sides are equal (SSS), the triangles are congruent.

14. (a) 314 (b) 1256

15. Replacing r by $2r$ in the area formula πr^2 gives $\pi (2r)^2 = 4\pi r^2 = 4A$. Multiplying the radius by any positive number k will have the effect of multiplying the area by a factor of k^2.

Chapter 5

1. (a) CXLV (b) CCXXXVII (c) MDCCCLXV (d) MCDXCII
   (e) MMCLXXXIX (f) MMMCMXCIX

2. (a) 2,259 (b) 2957 (c) 854 (d) 249 (e) 1,633 (f) 3,678

3. The symbol for 5,000 is $|\supset\supset$. The sum of (a) and (b) is $|\supset\supset$ CCXVI. The sum of (c) and (d) is MCIII.

4. To multiply, move the decimal point to the right the same number of places as there are zeros. To divide, move the decimal point to the left the same number of places as there are zeros.

7. (a) 2,881 (b) 1,794 (c) 20,076 (d) 6,993 (e) 61,632 (f) 393,669
   (g) 704,781 (h) 53,342,745

8. Consult a references such as Bibhutibhusan Datta and A. N. Singh, *History of Hindu Mathematics* (Asia Publishing House, 1962; reviewed: *Isis* 25, 478-488; reprint: Asia Publishing House, 1962); T. S. Bhanu Murthy, *A Modern Introduction to Ancient Indian Mathematics* (Wiley Eastern Ltd., New Delhi, 1992); or S. Balaachandra Rao, *Indian Mathematics and Astronomy* (Jnana Deep Publications, Bangalore, 1994).

9. (a) 六百七十八   (b) 五千八百一十二   (c) 九万八千七百六十一

(d) 一万二千三百四十五   (e) 二万三千七百八十九   (f) 二千二百一十二

10. (a) 4,355 (b) 976 (c) 8,692

12. $75\sqrt{3}$

14. 5

15. (a) $\frac{1}{4}$, 25% (b) $\frac{1}{2}$, 50% (c) $\frac{3}{4}$, 75% (d) $\frac{3}{10}$, 30%   (e) $\frac{1}{20}$, 5% (f) $\frac{1}{8}$, 12.5%   (g) $\frac{3}{2}$, 150% (h) $\frac{2}{1}$, 200%   (i) $\frac{1}{1000}$, 0.1% (j) $\frac{5}{4}$, 125%

Chapter 6

1. Let u_n = the number of pairs of rabbits on the nth month. The number of mature pairs of rabbits on the nth month is equal to the total number of pairs of rabbits the month before u_{n-1}. The number of nonmature pairs of rabbits is equal to the number of mature pairs of rabbits the month before which is equal to the total number of pairs of rabbits two months prior u_{n-2}. So the total number pairs of rabbits is the sum of the mature and nonmature pairs which can be expressed as $u_n = u_{n-1} + u_{n-2}$.

2. Write $u_1 = u_3 - u_2$, $u_2 = u_4 - u_3$, $u_3 = u_5 - u_4$, $u_4 = u_6 - u_5$, ..., $u_{n-1} = u_{n+1} - u_n$, $u_n = u_{n+2} - u_{n+1}$; adding all the equations we get $u_1 + u_2 + u_3 + u_4 + \ldots + u_n = (u_3 - u_2) + (u_4 - u_3) + (u_5 - u_4) + (u_6 - u_5) + \ldots + (u_{n+1} - u_n) + (u_{n+2} - u_{n+1})$. The sum on the right telescopes down to $u_{n+2} - u_2 = u_{n+2} - 1$.

3. (a) $3, 7, 11, 15, 19, 23, \ldots$ (b) $1, 4, 7, 10, 13, 16, \ldots$
   (c) $0, 2, 4, 6, 8, 10, \ldots$ (d) $5, 10, 15, 20, 25, 30, \ldots$
   (e) $7, 14, 21, 28, 35, 42, \ldots$ (f) $3, 8, 13, 18, 23, 28, \ldots$

4. The recurrence equation for an arithmetic sequence is $u_n = u_{n-1} + d$. Applying this rule again gives $u_n = (u_{n-2} + d) + d$. Again, $u_n = (u_{n-3} + d) + d + d$. Continuing in this manner gives $u_n = u_n - k + d + d + d + \ldots + d$ where there are k terms of d. Letting $k = n - 1$, we obtain $u_n = u_1 + (n - 1) \times d$.

5. (a) $1, 3, 9, 27, 81, 243, \ldots$ (b) $2, 6, 18, 54, 162, 486, \ldots$
   (c) $3, 6, 12, 24, 48, 96, \ldots$ (d) $1, \frac{1}{2}, \frac{1}{4}, 1/8, 1/16, 1/32, \ldots$
   (e) $1, -2, 4, -8, 16, -32, \ldots$ (f) $4, -2, 1, -\frac{1}{2}, \frac{1}{4}, -1/8, \ldots$

6. The recurrence equation for a geometric sequence is $u_n = u_{n-1} \times r$. Applying this rule again gives $u_n = (u_{n-2} \times r) \times r$. Again, $u_n = (u_{n-3} \times r) \times r \times r$. Continuing in this manner gives $u_n = u_{n-k} \times r \times r \times r \times \cdots \times r$ where there are k terms of r. Letting $k = n - 1$, we obtain $u_n = u_1 \times r^{n-1}$.

7. There are 13 lines: GGGGG, BGGGG, GBGGG, GGBGG, GGGBG, GGGGB, BGBGG, BGGBG, BGGGB, GBGBG, GBGGB, GGBGB, and BGBGB.

8. Starting with a and b, we get the following sequence: a, b, $a+b$, $a+2b$, $2a+3b$, $3a+5b$, $5a+8b$, $8a+13b$, $13a+21b$, and $21a+34b$. The sum of all of these is $55a+88b = 11 \times (5a+8b)$.

9. (a) $x = -6, x = 1$ (b) $x = -2$
    (c) $x = -3, x = 2$ (d) $x = (-3 \pm \sqrt{89})/4$
    (e) $x = \pm 10$ (f) $x = -1 \pm \sqrt{5}/5$
    (g) $x = -2, x = 5$ (h) $x = 6, x = -5$

10. Let $S_n = 1 + \frac{1}{3} + \frac{1}{9} + \frac{1}{27} + \cdots + \frac{1}{3^n}$. Then $3 \times S_n = 3 + 1 + \frac{1}{3} + \frac{1}{9} + \cdots + \frac{1}{3^{n-1}}$. Subtracting the two equations yields $2 \times S_n = 3 - \frac{1}{3^n}$. As $n \to \infty$, $\frac{1}{3^n} \to 0$. Thus, $2 \times S_n = 3$ or $S_n = \frac{3}{2}$.

11. Multiply each decimal by the appropriate power of 10, that is, 10^p where p is the number of digits in the pattern.

  (a) $\frac{8}{9}$
  (b) $\frac{23}{99}$
  (c) $\frac{450}{999} = \frac{50}{111}$
  (d) $\frac{9018}{9999} = \frac{1002}{1111}$
  (e) $\frac{54}{99} = \frac{6}{11}$
  (f) $2\frac{12}{99} = 2\frac{4}{33}$

12. $\frac{22}{7} = 3.\overline{142857}$ and $\frac{256}{81} = 3.16049382716\ldots$. Thus, $\frac{22}{7}$ is more accurate.

Chapter 7

2. Using the rule $x^n x^m = x^{n+m}$ and distributive law $a(b+c) = ab + ac$ for (e), (f), and (g), we have

  (a) x^6
  (b) x^4
  (c) x^8
  (d) $18x^9$
  (e) $40x^{10} - 20x^{13}$
  (f) $12x^{16} + 4x^{10} - 2x^7$
  (g) $4x^{13} + x^{17}$

3. (a) x (b) x^4 (c) $2x^7$ (d) $3x^3$ (e) $5x^2 + 3x^4$

4. Any nonzero number divided by itself is 1. Thus, $1 = \dfrac{x^n}{x^n} = x^{n-n} = x^0$.

5. Any number times 1 equals itself and 1 is the only number with this property. Since x^n times x^0 equals itself, it follows that $x^0 = 1$.

6. $\sqrt{x}$ is the positive number you square to get x, that is, $\sqrt{x}\sqrt{x} = x$, but $\frac{1}{2} + \frac{1}{2} = 1$ so $x^{\frac{1}{2}}x^{\frac{1}{2}} = x^1 = x$. Hence, $\sqrt{x} = x^{\frac{1}{2}}$.

7. Using the subtraction of powers rule, we have $\dfrac{x^3}{x^5} = x^{-2}$ while old-fashioned cancellation gives $\dfrac{1}{x^2}$. Therefore, $\dfrac{1}{x^2} = x^{-2}$. A similar argument works for x^{-n}.

8. Using $(x^m)^n = x^{m \times n}$ gives

   (a) x^{15}

   (b) x^{14}

   (c) x^{40}

   (d) x^{63}

   (e) $x^0 = 1$

   (f) x^2

   (g) $x^{\frac{2}{3}}$

   (h) x^{-10}

   (i) x^{-10}

Chapter 8

1. (a) The sum is 100. (b) Draw a 10-by-10 square array of dots similar to that in Figure 8-6.

2. The nth odd number is $2n - 1$. This formula, for example, states that the 10th odd number is $2 \times 10 - 1$ which is the correct answer of 19.

4. The tenth, eleventh, and twelfth triangular numbers are 55, 66, and 78, respectively.

5. Add the equations

$$
\begin{array}{rcccccccc}
t_n = & 1 & + & 2 & + & 3 & + \cdots + & & n \\
+\ t_n = & n & & + (n-1) & + (n-2) & & + \cdots + & & 1 \\
\hline
2t_n = & (n+1) & + (n+1) & + (n+1) & + \cdots + (n+1) &&&&
\end{array}
$$

so $2 \times t_n = n \times (n+1)$, which after dividing by two gives

$$
t_n = \frac{n(n+1)}{2}
$$

6. (b) $t_{n-1} + t_n = \dfrac{(n-1)n}{2} + \dfrac{n(n+1)}{2} = \dfrac{n^2-n}{2} + \dfrac{n^2+n}{2} =$
$\dfrac{n^2-n+n^2+n}{2} = \dfrac{2n^2}{2} = n^2$

7. Using some algebra, we have

$$
\begin{aligned}
t_n^2 - t_{n-1}^2 &= \left(\frac{n(n+1)}{2}\right)^2 - \left(\frac{n(n-1)}{2}\right)^2 \\
&= \frac{n^2(n+1)^2}{4} - \frac{n^2(n-1)^2}{4} \\
&= \frac{n^2(n^2+2n+1)}{4} - \frac{n^2(n^2-2n+1)}{4} \\
&= \frac{n^4+2n^3+n^2}{4} - \frac{n^4-2n^3+n^2}{4} \\
&= \frac{4n^3}{4} \\
&= n^3
\end{aligned}
$$

9. The first ten pentagonal numbers are 1, 5, 12, 22, 35, 51, 70, 92, 117, and 145.

Chapter 9

1. The following Java code can be used to determine if the integer *num* is a prime. Note that computers were not around in the time of Fermat.

```
/* A program to check primality. */
public class CheckPrime {
    public static void main(String[] args) {
        int num = 112303;
        int i;
        for(i=2; i<Math.sqrt(num); i++) {
            if (num%i==0) {
                System.out.println(num +
                    " is not prime. It has a factor of " + i + ".");
                break;
            }
            if (i>Math.sqrt(num))
                System.out.println( num + " is prime.");
        }
    }
}
```

4. (a) True (b) False (c) True (d) True (e) False (f) True

7. (a) 8 yards, 1 foot   (b) 4 yards, 6 inches
    (c) 4 yards, 2 feet, 10 inches   (d) 6 yards, 2 feet, 10 inches
    (e) 19 yards, 1 foot, 4 inches   (f) 27 yards, 2 feet, 4 inches

9. Since 123 mod 24 = 3, it will be 3 hours later or 6 P.M.

12. If one flips a coin 3 times in a row, there will be 8 outcomes, namely $HHH, HHT, HTH, HTT, THH, THT, TTH$, and TTT. Three of these have exactly one head (HTT, THT, TTH), making the probability $\frac{3}{8}$.

13. There are $3 \times 4 \times 6 \times 2 = 144$ possible complete dinners. Since the person only orders 1 of these, the chances are $\frac{1}{144}$ of guessing what they ordered.

17. (a) $y' = 12x + 6$ (b) $y' = 9x^2 - 10x + 2$
    (c) $y' = 8x^3 - 18x^2 + 20x - 4$ (d) $y' = m$

18. The antiderivative is x^3. The area is therefore $3^3 - 1^3 = 27 - 1 = 26$.

19. (a) Thursday (b) Wednesday (c) Thursday (d) Friday

20. There are $12! = 479{,}001{,}600$ possible permutations of the chromatic scale.

Chapter 10

1. (a) $V(G) = \{a, b, c, d, e, f\}, E(G) = \{ab, ac, bd, cd, ce, df, ef\}$, $\deg(a) = 2$, $\deg(b) = 2$, $\deg(c) = 3$, $\deg(d) = 3$, $\deg(e) = 2$ and $\deg(f) = 2$

   (b) $V(G) = \{a, b, c, d, e, f, g\}, E(G) = \{ab, bc, be, de, ef, eg\}$, $\deg(a) = 1$, $\deg(b) = 3$, $\deg(c) = 1$, $\deg(d) = 1$, $\deg(e) = 4, \deg(f) = 1$ and $\deg(g) = 1$

   (c) $V(G) = \{a, b, c, d, e, f, g, h\}, E(G) = \{ab, bc, cd, de, eh, ef, fg, gh, ha\}$, $\deg(a) = 2$, $\deg(b) = 2$, $\deg(c) = 2$, $\deg(d) = 2$, $\deg(e) = 3, \deg(f) = 2, \deg(g) = 2$ and $\deg(h) = 3$

4. When n is odd, the center of P_n consists of a single node. When n is even, the center of P_n consists of two adjacent nodes.

6. The radius and diameter of C_n is $\lfloor \frac{n}{2} \rfloor$, that is, n over 2 round down.

7. K_n is $(n-1)$-regular with $\dfrac{n(n-1)}{2}$ edges.

9. The total number of edges is $1 + 2 + 3 + \cdots + (n-1)$ which is t_{n-1}, the $(n-1)$st triangular number.

14. For a height of 200 ft., $V = 8\sqrt{200} = 80\sqrt{2}$. For a height of 400 ft., $V = 8\sqrt{400} = 160$.

15. $30 = 25 + 4 + 1$
     $31 = 25 + 4 + 1 + 1$
     $32 = 16 + 16$
     $33 = 25 + 4 + 4$
     $34 = 25 + 9$
     $35 = 25 + 9 + 1$
     $36 = 16 + 16 + 4$
     $37 = 16 + 16 + 4 + 1$
     $38 = 25 + 9 + 4$
     $39 = 25 + 9 + 4 + 1$
     $40 = 16 + 16 + 4 + 4$
     $41 = 25 + 16$
     $42 = 25 + 16 + 1$
     $43 = 25 + 16 + 1 + 1$
     $44 = 25 + 9 + 9 + 1$
     $45 = 36 + 9$
     $46 = 36 + 9 + 1$
     $47 = 36 + 9 + 1 + 1$
     $48 = 16 + 16 + 16$
     $49 = 49$
     $50 = 49 + 1$

Note: Some of the numbers above have more than one solution.

$$\begin{aligned}
100 &= 25 + 25 + 25 + 25 \\
&= 81 + 9 + 9 + 1 \\
&= 49 + 49 + 1 + 1 \\
&= 64 + 36 \\
&= 100
\end{aligned}$$

16. The seven partitions of 5 are 5, $4 + 1$, $3 + 2$, $3 + 1 + 1$, $2 + 2 + 1$, $2 + 1 + 1 + 1$, $1 + 1 + 1 + 1 + 1$. The partitions with three or fewer terms are 5, $4 + 1$, $3 + 2$, $3 + 1 + 1$, $2 + 2 + 1$. The partitions with no term greater than three are $3 + 2$, $3 + 1 + 1$, $2 + 2 + 1$, $2 + 1 + 1 + 1$, $1 + 1 + 1 + 1 + 1$.

Chapter 11

1. (a) $[0, 2, 0]$ (b) $[1, -4, 2]$ (c) $[-5, 6, -4]$ (d) $[5, -8, 6]$ (e) $[7, 0, 4]$
    (f) $[-5, 0, -2]$ (g) $[8, -6, 8]$

2. Consult the Thales proof in Chapter 3.

9. Pythagoras' Theorem states that $a^2 + b^2 = c^2$. Dividing both sides by c^2 gives $\dfrac{a^2}{c^2} + \dfrac{b^2}{c^2} = \dfrac{c^2}{c^2} = 1$ or $\left(\dfrac{a}{c}\right)^2 + \left(\dfrac{b}{c}\right)^2 = 1$. Substituting gives the desired result.

11. Using Pythagoras' Theorem, we obtain $c^2 = 1^2 + 1^2 = 2$. Thus, $c = \sqrt{2}$. Since the triangle is isosceles and right, the base angles are equal and their is sum is 90°. Hence, each one is 45°. It follows that the sine and cosine of 45° are both equal to $\dfrac{1}{\sqrt{2}} = \dfrac{\sqrt{2}}{2}$.

12. Both legs are $500\sqrt{2}$ ft.

Chapter 12

1. (a) 1001001   (b) 11101
   (c) 101110010   (d) 1100100000
   (e) 11111   (f) 10.1
   (g) 1100010.11   (h) 0.001

2. (a) 7   (b) 21   (c) 16
   (d) 54   (e) 3.5   (f) 31
   (g) 2.75   (h) 1.25   (i) 9.875

3. (a) 10001   (b) 110010
   (c) 111   (d) 10000000
   (e) 0.11   (f) 1.01
   (g) 1000   (h) 10000

4. (a) 1100   (b) 1001
   (c) 101   (d) 1001
   (e) 10101   (f) 0.001

5. (a) 1110010   (b) 1111110
   (c) 10101000   (d) 1110011

6. Placing a 0 at the end of a binary number has the effect of multiplying it by 2. Two zeros by $2^2 = 4$. Placing n zeros has the effect of multiplying the number by 2^n. A binary number is even if the last *bit* (binary digit) is 0. This last bit is referred to as the *parity bit*.

7. Shifting the decimal to the right n places has the effect of multiplying it by the power of two, 2^n.

8. 0.1111111... in binary is equal to 1.

9. (a) $2^3 3^1 5^3$ (b) $5^2 31^1$
   (c) *prime*   (d) $2^4 5^3 113^1$
   (e) $2^4 3^2 5^1$ (f) $2^{10} 3^{10}$
   (g) $2^2 5^2 11^1$ (h) *prime*

Chapter 13

1. (a) $1 \pm \sqrt{21}$
   (b) $-1 \pm \frac{\sqrt{6}}{2}$
   (c) ± 0.3
   (d) $0, \dfrac{1}{2}$
   (e) $\dfrac{1 \pm \sqrt{5}}{2}$
   (f) $-2, 6$

2. (a) $-2, 6$
   (b) 1
   (c) $2, 6$
   (d) $-1, \dfrac{1}{2}$
   (e) $-10, 4$
   (f) $-2, 12$

3. (a) axis of symmetry is $x = \dfrac{-(-4)}{2(1)} = \dfrac{4}{2} = 2$

   (b) axis of symmetry is $x = \dfrac{-(-2)}{2(1)} = \dfrac{2}{2} = 1$

   (c) axis of symmetry is $x = \dfrac{-(-8)}{2(1)} = \dfrac{8}{2} = 4$

   (d) axis of symmetry is $x = \dfrac{-(1)}{2(2)} = \dfrac{-1}{4} = -\dfrac{1}{4}$

   (e) axis of symmetry is $x = \dfrac{-(6)}{2(1)} = \dfrac{-6}{2} = -3$

   (f) axis of symmetry is $x = \dfrac{-(-10)}{2(1)} = \dfrac{10}{2} = 5$

4. (a) axis of symmetry is $x = 0$
    (b) axis of symmetry is $x = 0$
    (c) axis of symmetry is $x = \dfrac{-(-6)}{2\,(1)} = \dfrac{6}{2} = 3$
    (d) axis of symmetry is $x = \dfrac{-(8)}{2\,(2)} = \dfrac{-8}{4} = -2$
    (e) axis of symmetry is $y = 0$
    (f) axis of symmetry is $y = \dfrac{-(6)}{2\,(1)} = \dfrac{-6}{2} = -3$

5. (a) $x = 6$ and $y = 4$
    (b) $x = 4$ and $y = 1$
    (c) $x = 10$ and $y = 5$
    (d) $x = 3$ and $y = -2$
    (e) $x = 3$ and $y = -\dfrac{3}{2}$

6. (a) $x = 6$ and $y = 4$
    (b) $x = 4$ and $y = 1$
    (c) $x = 10$ and $y = 5$
    (d) $x = 3$ and $y = -2$
    (e) $x = 3$ and $y = -\dfrac{3}{2}$

7. (a) $\dfrac{x^2 - x + 1}{x^2 - 1}$
    (b) $-\dfrac{x^2 - 10x - 30}{x\,(x + 3)}$
    (c) $\dfrac{x^2 - 11}{x^2 - 5x + 6}$
    (d) $\dfrac{5x^2 + 5x + 2}{2x\,(x + 1)}$
    (e) $-\dfrac{x + 11}{(5x + 4)\,(3x - 1)}$

8. There are two solutions: $-12, -11$ and $11, 12$.

9. They intersect at $x = \dfrac{-1 + \sqrt{5}}{2}$ and $x = \dfrac{-1 - \sqrt{5}}{2}$.

10. They intersect at two points, $(1, 0)$ and $(0, -1)$.

11. They intersect at two points, $\left(\dfrac{\sqrt{2}}{2}, \dfrac{\sqrt{2}}{2} \right)$ and $\left(-\dfrac{\sqrt{2}}{2}, -\dfrac{\sqrt{2}}{2} \right)$.

12. The positive y value is $\dfrac{-1+\sqrt{5}}{2}$.

The two x values are $\pm\sqrt{\dfrac{-1+\sqrt{5}}{2}}$.

Index